IMPACT OF MACHINE LEARNING IN DIFFERENT SECTORS

IMPACT OF MACHINE LEARNING IN DIFFERENT SECTORS

Dr. J.W. Bakal
Dr. Ashish Tripathi
Dr. Mansi Subhedar
Prof. (Dr.) B.K. Sarkar

NAMYA PRESS.com

Published by

Namya Press

USA: 10685-B Hazlehurst Dr. # 31013 Houston, Texas - 77043
Email: namyapress@gmail.com
Website: https://namyapress.com

Edition	:	First Published in 2022 (Namya Press)
Title	:	Impact of Machine Learning in Different Sectors
Editors	:	Dr. J.W. Bakal, Dr. Ashish Tripathi, Dr. Mansi Subhedar and Prof. (Dr.) B.K. Sarkar
ISBN	:	978-1-956861-13-6
Copyright	:	© Dr. J.W. Bakal, Dr. Ashish Tripathi, Dr. Mansi Subhedar and Prof. (Dr.) B.K. Sarkar. 2022 All rights reserved

Preface

It's with great happiness that I would like to acknowledge a great deal of people that get helped me extremely through the entire difficult, challenging, but a rewarding and interesting path towards some sort of Edited Book Without having their help and support, none of this work could have been possible.

Dr. J.W. Bakal, Dr. Ashish Tripathi, Dr. Mansi Subhedar, Prof (Dr.) B.K. Sarkar highly indebted to Dr. Vasudevan Pillai, Secretary and CEO, and Dr. Daphne Pillai, Joint Secretary and Chairperson, Management Board of Mahatma Education Society, for their kind support and encouragement.

Contents

Chapter - 01

Sentiment Analysis Using Deep Learning

Harsh Patil

MES's, Pillai HOC College of Engineering and Technology, Rasayani, Maharashtra India

Sonia Koli

MES's, Pillai HOC College of Engineering and Technology, Rasayani, Maharashtra India

Dr. Mansi Subhedar

MES's, Pillai HOC College of Engineering and Technology, Rasayani, Maharashtra India

ABSTRACT

In today's technology, human-machine interaction is becoming more common, and robots must be able to comprehend human actions and reactions. Human behavior can be better understood if technology can recognize person's emotions, resulting in increased work efficiency. Text, audio, linguistic, and facial expressions can all be used to communicate emotions. Facial expressions are important in judging a person's emotions. In this work, an approach for classifying human moods using face pictures is presented using deep learning technique. The suggested approach uses Haar-cascade face recognition, Active shape Model (ASM) data retrieval (26 face points collected), and Ada-boost algorithm to categorize five emotional responses: rage, hate, pleased, neutrality, and amazement. The proposed approach is unique in that it does the emotional expression in actual environments on the Raspberry Pi, with a real-time average accuracy of 94%. In social surroundings where sentiment identification is vital, the Raspberry Pi can recognize sentiments flexibly in live time when it has been joined to a robotic system.

INTRODUCTION

In today's innovation, collaboration between people and machines is becoming ever more widespread, as well as robots must be able to comprehend human actions and reactions. If one system can understand people's emotions, it will be able to better interpret individual interactions, which will lead to higher productivity. It might be used as a critical assessment tool in behavioral research, and socially skilled robot software could be developed. Emotions are strong sentiments affected by the environment that play an important role in daily activities including judgement, training, concentration, inspiration, adapting, perceiving, organizing, cognitive, understanding, and several others, making emotion detection a large study topic. Text, auditory, linguistic, and visual expressions may all be used to recognize emotions. According to [1,] linguistic signals constitute 7% of communication in human-to-human interactions and facial emotions contribute the majority of the remaining 55%. As a result, examination of facial movements is amongst the far more crucial elements in sentiment categorization. Although facial expression detection from two-dimensional images is a very well-observed topic, there is currently no way for estimating traits in moderate quality pictures in actual environments. The majority of these studies [2]-[3] is founded on frontal view images of people. Because these global settings are not uniform in real life, additional research on non-frontal images with varying illumination conditions is required. Using the CMU multi-pie database [4], which comprises of 2D photos with varying brightness and locations, we present a sustainable emotion recognition system that recognizes fundamental feelings such as anxiety, contempt, joy, astonishment, and neutrality. Because of its small size, lightweight, and low power supply needs, the computer program created following our recommended approach is installed on a Raspberry Pi, which may be used with robots. As a consequence, it may be easily installed on any robot and used for a variety of reasons, such as security systems, supervising senior people or kids at home, observing complicated cases in Emergency departments, client satisfaction, and so on. The Raspberry Pi circuit board is shown in Figure 1. The following is a breakdown of the paper's structure: Part II discusses background research, section III discusses the recommended strategy, and section IV discusses the findings and analysis. A conclusion and ideas for further improvement are included in the last section.

The objective is to identify the perfect option for real-time sentiment identification in online education systems' face detection and recognition. It is essential to maximize the performance and reliability of computer vision algorithms in order to accomplish this aim.

Related Work

A major element of social connection is the consistency of facial emotions and mannerisms. Charles Darwin remarked on the worldwide widespread facial features throughout the eighteenth millennia, that are important during unspoken information exchange [5]. In 1971, Ekman et al described maxillofacial features as generally connected with diverse emotions [6]. Both humans and animals appear to make discrete muscle motions that correlate to a certain mental state. Feature extraction is usually followed by classification in an image classification system. Fasel et al presented a comprehensive overview of quantitative learning algorithms and classification techniques for identifying facial emotions [8]. Considering the prevailing accessibility of data sets and processing capability, however, deep neural network performance is projected to be greatly improved. Some recent milestones are listed here. Krizhevsky et al discussed automated image classification [9]. This study shows that a deep neural network with capabilities equal to the human sensory system may be developed. A model to detect things from photos was built by using the CIFAR-10 database and a self-made named collection of 60,000 photographs sorted into 10 groups. A further significant outcome of the research is the visualization of the filters for the internet, this enables the algorithm to deconstruct and examine pictures. The creation of the yearly Image net challenges [10] in 2010 sparked interest in photo categorization, and the huge amount of labeled data that resulted has been regularly used in papers since then. According to Krizhevsky et al. [11], the ImageNet LSVRC-2010 challenge teaches a structure of Five convolutional, from 1.2 million moderately lifted pictures, there are three collecting levels and three completely linked regions. Lv et al. [12] propose a deep facial emotion recognition network that works well with JAFFE and larger CohnKanade (CK+) libraries. On the same database, alternative techniques including Support Vector Machine (SVM) and Learning Vector Quantization (LVQ) achieve an accuracy of around 95%.

The current dataset is the Facial Emotion Recognition Competition (FERC-2013) [13], which achieved an average emotion classification accuracy of 67.02 percent using an ordered deep network.

Proposed Work

Artificial intelligence and machine learning and cognitive computing (Automation) technology thus mimic the way humans comprehend information. Deep learning is a major component of data science, which includes statistical and predictive analysis. Deep learning is especially beneficial to data professionals who are responsible for obtaining, evaluating, and analyzing big quantities of information just because it simplifies and enhances productivity [14]. Machine learning algorithms may very well be regarded about it as a means to systematize advanced analytics at even the most fundamental level. Artificial neural networks, with the exception of common data mining metaheuristics, appear to be straightforward and therefore are constructed in a scalability and conceptualization hierarchy. Deep learning has the advantage of creating the set of features by itself, without the use of human participation.

Intelligent machines involve large amounts of information, data, and computational power to attain a reasonable degree of precision, but none of it was broadly available to designers just before the advent of big data as well as cloud-based applications. Although it can immediately construct complex computer simulations through its own recurring results, deep learning programming may develop appropriate forecasting models from massive volumes of unlabeled, unstructured data. As the internet of things (IoT) grows increasingly popular, this is crucial since most of the data people and technology acquire is unorganized and unclassified.

In this study, we employ a robust CNN to identify pictures in order to make use of Google's deep learning (DL) public library "Keras" in automatic facial expression recognition. We used two distinct datasets to train the suggested network and two separate databases to test its validation and loss accuracy. We employed a sentiment model created by a CNN utilizing deep learning to identify sentiments in photos taken from a dataset that included seven distinct facial expressions. In comparison to the previous approach, which was utilized on some kind of Google-provided Keras library, we altered a few phases in CNN, as well as the CNN architecture, to enhance accuracy. The suggested structure was designed leveraging Keras for sentiment recognition.

Convolutional Neural Network (CNN) Architecture

A convolutional neural network (CNN, or Conv-Net) [15] seems to be a form of Artificial Neural Network (ANN) that is used in deep learning to analyze visual pictures. Shifting Invariant or Spatial Invertible Artificial Neural Networks (SIANN) are determined by the perceived architecture of convolutional or strainers which move over attribute values and generate translation equivariant outputs referred to as data maps. Interestingly, rather than being consistent during transcription, the majority of deep neural networks are merely equivariant. Only a few of the applications include multimedia content identification, recommendation systems, image classifications, pattern classification, information retrieval, computational linguistics, central nervous system interfaces, and monetary time series.

In CNNs, multilayer perceptrons (also known as artificial neurons) are regularized. Multilayer perceptrons are networks wherein every neuro inside one level is linked among all neurons in the following layer. Because of their “full connectivity,” such networks are susceptible to data errors. Normalization, or the prevention of errors, may be achieved in a number of ways, involving compensating factors throughout conditioning (like losing weight) or limiting interconnectivity. CNNs adopt a different method of generalization: those that utilize the advantage of the hierarchical structure in data and generate designs of growing difficulty using smaller and more manageable sequences imprinted in respective detectors. As an outcome, CNNs seem to be at the bottom of the interconnectivity and difficulty scale.

The convolution network connection layout was a biologically inspired phenomenon since it resembles the organization of the creature’s visual cortex. Part of the theoretical neurons only responds to stimuli inside the region of interest, a tiny area of the field of view. The convolutions of sensory neurons coincide, permitting it to cover the entire peripheral vision.

CNNs need extremely little pre-processing compared to other image categorization algorithms. The above appears to mean that, in contrast to traditional techniques, the infrastructure understands to improve the scrubbers (or polynomial kernel) through feature learning. A notable advantage is the lack of dependency on previous data or human intervention in extracting the features. The winning approach included a mechanism for detecting mood based on facial characteristics. However, cropping the ocular area and taking the specific distribution is the sole choice during the preprocessing stage, resulting in poor emotion recognition results. As a result, this procedure has to be tweaked to improve accuracy. Figure 2 shows the proposal solving approach in detail. Face features can be added via Haar-Cascades to obtain relevant information.

To do so, we employ the Haar-Cascades approach, to determine if the photographs have a face, and if they don’t, we go back to the beginning and enter the picture frames. If indeed the visage is present, the eye and lips must be located and the regions around the eyes and mouth trimmed. The Sobel edge detection procedure is used to identify and find edges, proceeded through image retrieval. To teach the evacuation of features, we employ a neural network technique. To retrain the computational model, we used a range of relevant pictures from six different sentiment classifications. The intake patterns are passed across all layers of unit operations during the first process, generating an external to the process in proportion to the training data. The mistakes found throughout the output variables are transmitted to the hidden neurons in the second stage, where the connection parameters typically modified to lessen overall defect. The above technique is performed till the outcome relative fault for all segments in the training examples falls below a certain threshold.

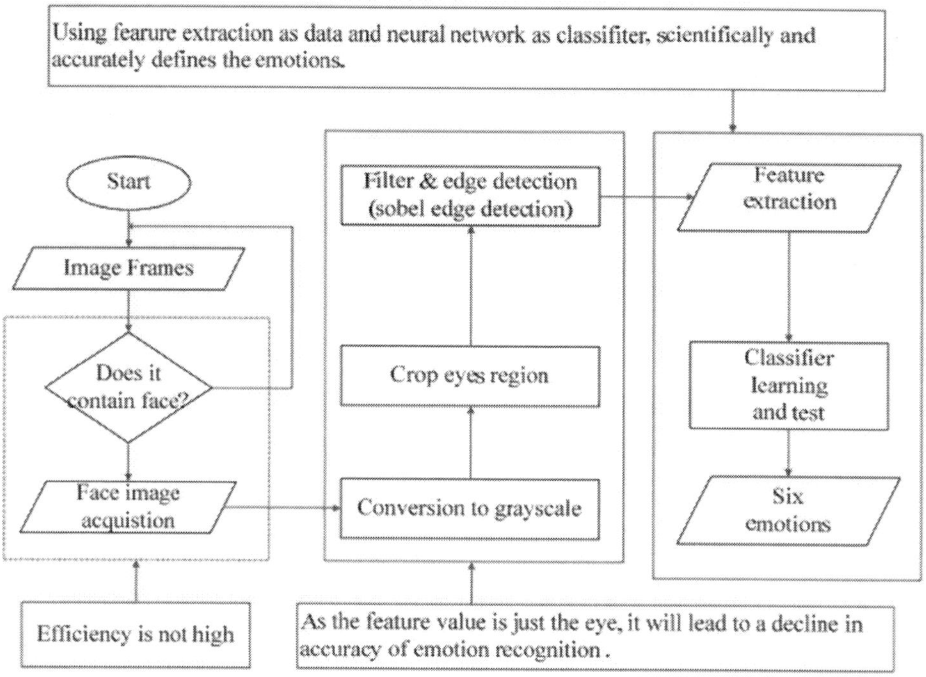

Fig..1. Current best solution

A neural network is built using the optimum range of human vision. The structure is fed with the best values from the pre-processed data. One input neuron, single convolution layers of 20 synapses, and seven outputting neurons make up the network. The estimator's independent and dependent variables are individually contradictory binary nibbles that indicate a mood.

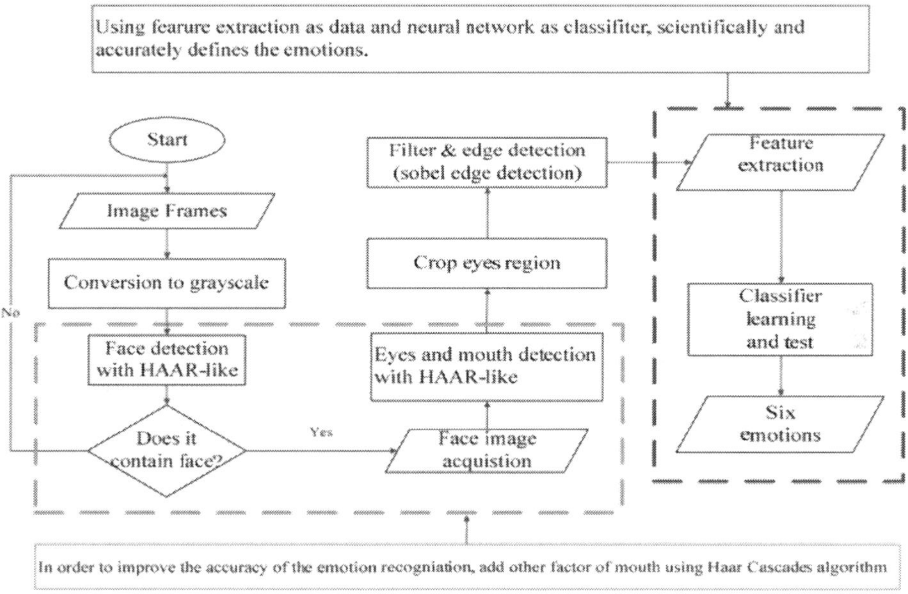

Fig. 2. Proposed solution

Haar Cascades are used to locate eyeballs and mouths. Velten as well as Sablik et al [16] utilize entity cascading classification. Haar - like features identification in their effective object identification approach, which is inspired by Paul Viola as well as Michael Jones' research "Rapid Object Detection using a Boosted Cascade of Simple Features." It's a machine learning approach that uses a huge number of picture training cascade functions, both favorable and unfavorable. This can then be used to identify objects in plenty of other photographs. Again, for the suggested protocol, we first trained raw photographs, recover facial parameters, and save them in the file face.xml. The eyes region and mouth are recognized based on the characteristics of the face. This mechanism for recognizing eyes and mouth is detailed further down.

Algorithm : Eyes and mouth detection INPUT: Test image samples images (I) = {I1, I2, ...,In}

OUTPUT: Detected eyes and mouth BEGIN

Step 1: Input the samples images

(I) = { I1, I2, ..., In }

Step 2: Read the image and store to the frame F = { I1, I2, ..., In }

Step 3: Read the contents of the xml file for face detection and store in the memory

R = face.xml //read face.xml, it stores in OpenCV lib. Step 4:

i= {1,2,3,...,n}

fi ={f1, f2,...fn} START loop F0 = 1

i = 0

Step 5:

While Fi > F(target) and i < n(Stages) i = i + 1

Train classifier for stage(i) Initialize weights Normalize weights

Pick the (next) best weak classifier Update weights

Evaluate Fi if Fi > f

go to Normalize Weights

Combine weak classifiers to form the strong classifier stage

compute Fi

Step 6: End of the algorithm

Filter & edge detection

Average and middle filtration were employed for smoothing out the picture by removing any previously existing unwanted noise. The border is generated at the place where another variation inside the supplied photos is at its maximum. These details of equation 2 for eye sockets and lips edge are shown below.

The main aim of the proposed technique seems to be to improve actual environment emotion recognition through image features with the purpose of distinguishing basic emotions including aggression, hatred, pleasure, amazement as well as neutrality.

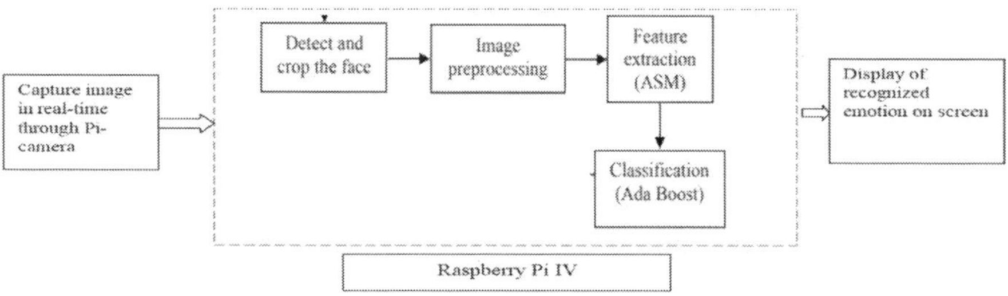

Fig. 3: Block diagram of real-time emotion recognition system

A 1.5 Gigahertz 64-bit 4x4 ARM Cortex-A72 CPU, across 802.11ac Area network, GPS/ a - GPS 5, filled gigabit (utilization not restricted), 2 different Standard USB connectors, two USB connectors, 1–8 Gigabytes, and dual-monitor allows for up of about 4K resolution through a duo of mini-HDMI (HDMI Type D) slots are all included in the Raspberry Pi 4 Model B. The Pi 4 can also be charged via a USB-C port when connected to a sufficient power source, enabling greater power to be provided to downstream devices. The Pi, unlike other small devices in this class, can only run on five volts, not Nine or Twelve volts.

Figure 2 depicts the proposed system's design, which is spelled out as described in the following: The raw image is captured in actual time by the camera and then sent into the emotion detecting program as input. On the Raspberry Pi, emotion identification software is installed, and the outcome is an identified emotion. The identified sentiment is presented on the computer screen. The undertakings carried out from these programs loaded on the processor which is Raspberry Pi as depicted in Figure 3.

The following is the method for allowing actual time emotional expressions upon that Raspberry Pi:

Step 1: A webcam is being used to capture the input image.

Step 2: The Viola-Jones [17] The biometric templates is detected using an image processing technology. Using the Haar wavelet principle, Viola-Jones created an integrated image to recognize faces. Because each portion of the visage seems to have a unique intensity value, Haar features account for the fact the magnitude of readings in surrounding adjacent pixels. After detection, the biometric templates are preserved for further computation, while the non-facial region gets eliminated.

Step 3: Cropping the picture to the required size and turning it into a grey image are the final steps in image processing. The noise in this cropped image is smoothed out with the Sobel filter.

Step 4: For extracting the features, the Active Shape Model (ASM) is used, which would be founded on a geometrical methodology. After using the ASM automated correlation-based spot localization algorithm to a facial expression image, Euclidean measurements are seen between centered gravitational position, and also the documented calibration position dimensions of the input images are calculated.

26 face points are retrieved and Euclidean distances in them are computed during real-time testing, as shown in Fig. The length here between the middle of the brow and the eyeball in the right eye is measured using points 1, 3, 5, and 7. The opposite eyes, as well as different locations' Euclidean distances, are then so same method of calculation, as indicated in Table I, using the equation (1)

$$D = \sqrt{(xi - xj)2 + (yi - yj)2} \qquad ...(1)$$

x, y seem to be the retrieved point is obviously parameters in the two-dimensional picture, and I j = 1, 2, 3... 26.

As indicated in Table I, ten face features are produced by combining 26 retrieved biometric information. Because the 10 facial traits exhibit varied relative importance of various reactions, the quantities of the physical expressions are preserved and compared.

Step 5: The data is classified using an adaptive boosting classifier (AdaBoost). AdaBoost is indeed a revolutionary educational paradigm that helps people learn supervised classification. That integrates the achievement of several classification algorithms to generate a formidable panel, as it is being shown in equation (2) [18]. AdaBoost is indeed a versatile advertising classification algorithm that works with almost any learning approach. It's a simple and uncomplicated process that just changes a single component, a number of repetitions, to obtain great accuracy.

$$(x) = \sum_{t=1}^{r} ft(x)$$

After the amount of repetitions T, f(x) become feeble classifiers and F(x) is powerful classifiers.

Table 1: Pairing of Extracted Points to Find Facial Feature

Facial feature	Paired extracted points
Eyes	1, 2, 3, 4, 5, 6, 7, 8
Forehead middle	11, 12,
Forehead + Chin	9, 10, 16, 17
Cheek1	13, 21
Cheek2	14, 23
Cheek3	20, 22
Cheek4	24, 19
Lip height	25, 26
Lip width	Lip width
jawline	15, 18

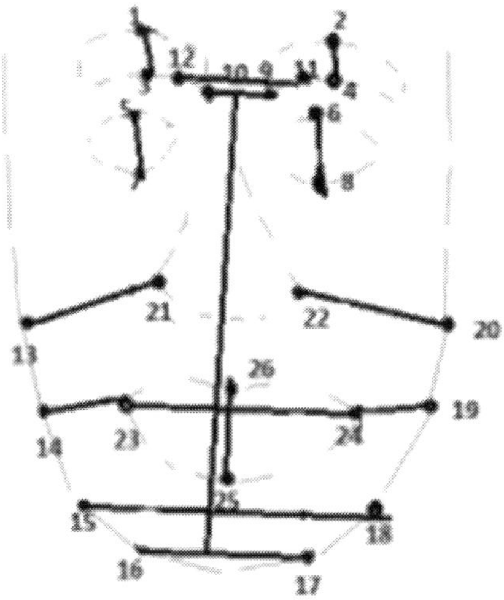

Fiducial points used in the proposed work

Step 6: Raspberry Pi design and implementation: The actual real application is evaluated and implemented on a Raspberry Pi in a Linux environment. Figure 7 shows the proposed system architecture, which includes a Raspberry Pi, an auxiliary camera, keypad, and display monitor.

Because the Raspberry Pi lacks a window or input device, it is linked to a display screen and keyboard. Using Virtual Network Connection (VNC) as well as putty software, a laptop may be utilized as a virtual computer for viewing and input. A person's image is collected and transferred to the Raspberry Pi when they gaze into the webcam in real time. Emotions will be identified and shown on the display panel using previously installed emotion recognition software.

Result and Discussion

The suggested emotion identification system based on face recognition is put to the test in this part. This approach employs a Python 3 platform and the OpenCV 2.0 library. Image transactions are simple to handle because of OpenCV 2.0's extensive functionality. To tackle the computer vision problem, the program imports the OpenCV library. The OpenCV library function is primarily used to identify and track the object's joint data.

The photographs in this study were tested using the JAFFE repository. There are 213 photos of Japanese people in the collection of people's faces (every picture is 256 * 256 pixels), each with its own distinct sentiment. In the expressive library, there are ten people. This input image was being used to adjust the cut, putting the eyeballs inside the repository picture in nearly the same spot as that of the actual picture; facial measures are virtually equal. Because this expression database is generally accessible and indeed the sentiment measurement is consistent, the majority of editorials on sentiment identification as well as recognition research have been used to test the proposed.

Table 2 compares all the effectiveness of the suggested and current solutions, showing the inaccuracies and inefficiencies. This evaluation consists of the following sets of data, each of which can be used to express six distinct emotions (enjoyment, grief, amazement, wrath, disdain, and fright), with the final set of statistics presenting just joy and sorrow expressions. A combination of 20 images was utilized, with efficiency evaluated in seconds and consistency assessed in percentages. More information can be found in Table 4. Table 2 displays the approximate correctness for the six major sentiments, with Table 3 providing additional information. Table 3 reveals that happiness was detected at a significantly higher rate than the other emotions, with surprises coming in second and grief coming in third. The determining factors of joyous, as well as the representative values of certain other facial features, is readily apparent, however, the determining factors of sadness, which frequently include other emotion identification key characteristics, are less so, resulting in poor prediction performance.

Table 2 displays the approximate time-span consumption (second) for every sentiment, whereas Table 4 provides further information. Overall, regardless of whether the optimal or suggested solution is employed, the processing time required is rather considerable; this is far from commercial use. In terms of overall development time, the suggested strategy is more efficient than the optimal choice. Surprisingly, surprise takes the longest to recognize, accompanied by fright, enjoyment takes the shortest, while frustration and anger take approximately the same amount of period to recognize.

Furthermore, accuracy has improved once the mouth was updated as a distinguishing element using a Harr-style methodology. The findings show that the suggested method is superior to the current one.

Algorithm 2: eyes and mouth edge

INPUT: Detected image(I) OUTPUT: Edged image(I)

BEGIN X={1,2,3,...n}

C is that point the image in the frame S is that sotre the image

X=0, C = 0, S=0

Step 1: Read the detected image from the system memory

R = Detected image(I)

Step 2: Detect in the image START loop

For X = 1 to n

Read each pixel in In

Point the image in the frame

C = Sobel (In) // import sobel function in OpenCV lib Store the image in the system directory

S = Store(C) // import store function in // OpenCV lib End of For loop

Step 3: End of the algorithm 2

Table 2: Comparison Results

Sample images	**Proposed solution (Haar cascade for eyes and mouth)**				**Current solution (Sobel edge for eyes detection)**		
	Output	**Detected emotion**	**Accuracy (%)**	**Processing time**	**Output**	**Accuracy**	**Processing time**
	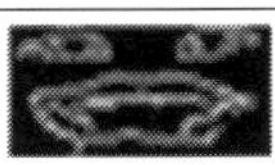	Sad	81	135.01		78	141
	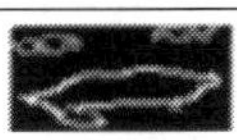	Angry	88.5	110.5		81	120
	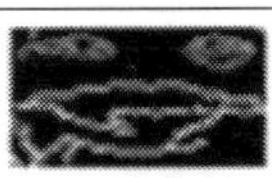	Happy	93	114.5		89.5	121
		Surprise	90.8	163		88	171
		Neutral	80	165		70	175

Table 3: Average accuracy (%) of each emotion

	Sad	Surprise	Happy	Angry	Neutral
Proposed Solution	78.54	93.26	95.25	90.22	80
Current Solution	76	87.2	87.6	82.7	79.3

Table 4: Average processing time (sec) of each emotion

	Sad	Surprise	Happy	Angry	Neutral
Proposed Solution	130.27	153.7	110.3	112.60	100
Current Solution	143.7	170.5	121	116	120

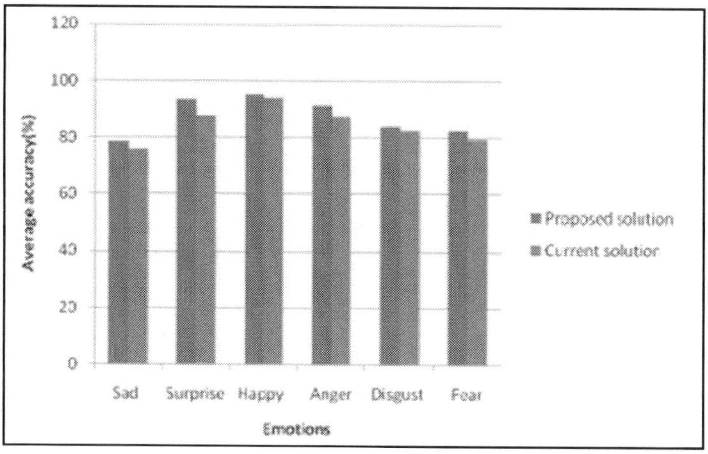

Fig. 3: Average accuracy (%) of each emotion

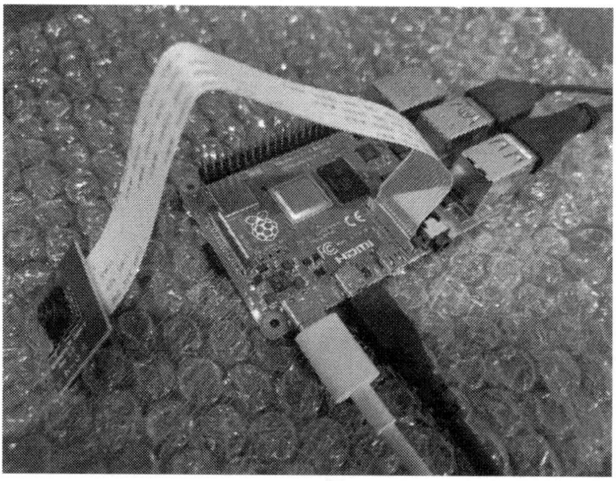

Fig. 4: Raspberry pi connected with pi camera module

Fig. 5: Sample output for neutral emotion

Fig. 6: Sample output for Happy emotion

CONCLUSION

This paper provides a facial recognition-based technique for dealing with the challenges of emotion classification in digital realities, while simultaneously addressing precision and effectiveness. Applying the HAAR Cascades algorithm to recognize face and eyes and distinguish all the sorts of sentiments using a machine learning approach, a mixture of efficiency and reliability is achieved. It would have the potential to be utilized for remote learning in the real world. The issue of emotion identification in digital realities has gotten a lot of interest. As a result of the shift in uncertainty factors, teachers and pupils are confronted with a more intricate pattern.

As a result, identifying emotions in an online learning network application mode is challenging. As a result, it is recommended that the problem be investigated further in the following areas:

Because the lighting and location of the image were not taken into account in this investigation, it's unknown how much these factors impact facial features as well as, as a result, emotion detection. All of these issues will require further investigation and testing in the future.

It is proposed that image enhancement, neural networks, cognitive science and pattern recognition, psychology, computer vision, and integration with other biometric identification models and approaches of collaborative and interactive perspective be used to fully meet the practical requirement, based on extensive and meticulous research.

Different algorithms might be used in the future to increase recognition accuracy. Neurological inspiration may also be used to teach robots to identify emotion. For emotion recognition, other modalities such as audio can be integrated with picture.

REFERENCES

[1] Mehrabian, A. Communication without words, Psychology Today, volume 2, pp 52-55, 1968.

[2] Peng Zhao-yi, Zhu Yan-hui, Zhou Yu, Real-time Facial Expression Recognition Based on Adaptive Canny Operator Edge Detection, IEEE, Multimedia and Information Technology (MMIT), Second International Conference on (Volume:2), pp 154 – 157, 2010.

[3] Michel F. Valstar, Member, IEEE, And Maja Pantic, Senior Member, IEEE, Fully Automatic Recognition Of The Temporal Phases Of Facial Actions, IEEE Transactions On Systems, Man,And Cybernetics-Part B: Cybernetics, Vol. 42, pp 28–43,2012

[4] Ralph Gross, Iain Matthews, Jeffrey Cohn, Takeo Kanade, and Simon Baker," Guide to the CMU Multi-PIE Face Database", Robotics Institute, Carnegie Mellon University.

[5] C. R. Darwin. The expression of the emotions in man and animals. John Murray, London, 1872.

[6] P. Ekman and W. V. Friesen. Constants across cultures in the face and emotion. Journal of personality and social psychology, 17(2):124, 1971.

[7] J. Nicholson, K. Takahashi, and R. Nakatsu. Emotion recognition in speech using neural networks. Neural computing applications, 9(4): 290–296, 2000.

[8] B. Fasel and J. Luettin. Automatic facial expression analysis: a survey. Pattern recognition, 36(1):259–275, 2003.

[9] A. Krizhevsky and G. Hinton. Learning multiple layers of features from tiny images, 2009.

[10] J. Deng, W. Dong, R. Socher, L.-J. Li, K. Li, and L. Fei-Fei. Imagenet: A large-scale hierarchical image database. IEEE Conference on Computer Vision and Pattern Recognition pages 248–255.IEEE, 2009.

[11] A. Krizhevsky, I. Sutskever, and G. E. Hinton. Imagenet classification with deep convolutional neural networks. In Advances in neural information processing systems, pages 1097–1105, 2012.

[12] Y. Lv, Z. Feng, and C. Xu. Facial expression recognition via deep learning. In Smart Computing (SMARTCOMP), 2014 International Conference on, pages 303–308. IEEE, 2014.

[13] E. Correa, A. Jonker,M.Ozo, andR.Stolk, "Emotion recognition using deep convolutional neural networks," Tech. Report IN4015, 2016.

[14] Jsj Ren and L. Xu, "On Vectorization of Deep Convolutional Neural Networks for Vision Tasks", AAAI, 2015.

[15] Shrestha A, Mahmood A. Review of deep learning algorithms and architectures. IEEE Access. 2019;7:53040–65.

[16] Sablik, Velten, Kummert (2013), "Automatic Privacy Protection Based on Haar Cascades and Spectral Domain Watermarking", VDE VERLAG GMBH, ISBN: 978-3- 8007-3543-3, pp 199-204.

[17] Paul V., Michael J., 'Robust Real-time Face Detection',International Journal of computer vision, May 2004, Volume 57, Issue 2, pp 137-154

[18] Yoav Freund. An adaptive version of the boost by the majority algorithm. In Proceedings of the Twelfth Annual Conference on Computational Learning Theory, 1999

CITATIONS

- Dr. Reni Francis ,2021107208, Granted, MAHATMA EDUCATION SOCIETY'S ," Analyzing the Role of Edu-2030 Model in Developing Competencies with respect to SDGs among Student Teachers" ,2021, Australia Patent Office.
- Francis, Reni; Cherian, Jaya ,2021105656, Granted, MAHATMA EDUCATION SOCIETY'S "Analysing the role of SHANTI Model of Teaching in fostering Peace among Secondary School Students" 2021, Australia Patent Office.
- Mr. Raju Narwade, Mr. Aditya Jadhav, Mr. Varad Salunkhe, Mr. Atharva Ugale, Ms. Smitha J, 202121039462 ,MAHATMA EDUCATION SOCIETY'S PILLAI HOC COLLEGE OF ENGINEERING AND TECHNOLOGY," Method and Process of Manufacture P-Tiles using Waste Plastic and other Required Material" 2021,Indian Patent office.
- Dr. Lata Menon, Lalit Mahato, Yash Khopkar, Priya Sharma, Ms. Darshana Wajekar, Mr. Ravi Bari, 202121039911,MAHATMA EDUCATION SOCIETY'S PILLAI HOC COLLEGE OF ARTS, SCIENCE AND COMMERCE "FLSA-System: Fifth-level Authentication System (Mobile, Laptop, PC, Locker, Etc.)" 2021, Indian Patent office.
- Dr. Archana Bhagwat, Ms. Krutika Mahendra Malkar, Mr. Abhishek Jeevan Ingole, Ms. Srushti Satyawan Sawant, Dr. Vishakha Telgote, 202121040919, MAHATMA EDUCATION SOCIETY'S PILLAI HOC COLLEGE OF ARTS, SCIENCE AND COMMERCE, "Intelligent Process and Methods for Casein Determination in Cow-milk only" 2021, Indian Patent office.
- Dr. G. V. Patil , Sagar Tate, Pranav Gawand, Gaurav Patil, Mr. Hemant Patil, Ms. Ashwini Kadam, 202121042117, MAHATMA EDUCATION SOCIETY'S PILLAI HOC COLLEGE OF ENGINEERING AND TECHNOLOGY, "Automatic Air Fill in Bike, Scooty and Car Tyres" 2021, Indian Patent office.
- Dr. Biplab Kumar Sarkar, US-17463679, MAHATMA EDUCATION SOCIETY'S PILLAI HOC COLLEGE OF ENGINEERING AND TECHNOLOGY "Advanced Kaplan Turbine." 2021, US-Patent Office.

Chapter - 02

Product Ranking using Machine Learning

Rutuja Vivek Tikait

Assistant Professor, Pune University.

Dr. G.R. Patil

Mahatma Education Society's, Pillai Hoc College of Engineering & Technology Rasayani
Taluka Panvel, Dist, Navi Mumbai, Maharashtra 410207, India.

Mr. Raju Narwade

Mahatma Education Society's, Pillai Hoc College of Engineering &Technology Rasayani
Taluka Panvel, Dist, Navi Mumbai, Maharashtra 410207, India.

ABSTRACT

Our Invention "Product Ranking using Machine Learning" is a An item might have many angles. A portion of the item angles are a higher priority than the others and have solid effect on the inevitable customer's decision making as well as association's item advancement systems. Recognizable proof of significant item perspectives become vital as the two buyers and firms are benefited by this. Buyers can without much of a stretch settle on buying choice by focusing on the significant viewpoints also as firms can zero in on working on the nature of these perspectives and in this manner upgrade item notoriety productively. This paper gives the depiction of different methods for item angle ID and grouping

BACKGROUND OF THE INVENTION

By and large, an item might have number of viewpoints. For instance, a Smart Phone has many viewpoints, for example, "screen size," "camera," "memory size," "sound quality." one might say that a few perspectives are a higher priority than the others, and have solid effect on the purchasers' decision making as well as firms' item improvement systems.

For instance, a few parts of Smart Phone e.g., "camera" and "memory size," are thought of as significant by a large portion of the purchasers, and are a higher priority than the others, for example, "shading" and "fastens.

Thus, the recognizable proof of significant item perspectives assumes a fundamental part in working on the convenience of audits which is valuable to the two shoppers and firms. Buyers can without much of a stretch settle on buying choice by focusing on the significant viewpoints, while firms can zero in on the improvement of item quality so item notoriety is upgraded.

Nonetheless, manual recognizable proof of significant perspectives is illogical. Along these lines, a way to deal with consequently distinguish the significant perspectives is exceptionally requested. Spurred by the above perceptions, we made a study on various strategies used to find significant item perspectives consequently from online buyer audits.

OBJECTIVES

The objective of the invention is to provide a "Product Ranking using Machine Learning" is a An item might have many angles. A portion of the item angles are a higher priority than the others and have solid effect on the inevitable customer's decision making as well as association's item advancement systems.

The other objective of the invention is to provide a Recognizable proof of significant item perspectives become vital as the two buyers and firms are benefited by this. Buyers can without much of a stretch settle on buying choice by focusing on the significant viewpoints also as firms can zero in on working.

The other objective of the invention is to provide a nature of these perspectives and in this manner upgrade item notoriety productively. This paper gives the depiction of different methods for item angle ID and grouping

SUMMARY OF THE INVENTION

The course of item viewpoint positioning comprising of three fundamental Steps: (a) angle ID; (b) opinion arrangement on perspectives (c) Product angle positioning. Given the shopper audits of an item, first recognize the angles in quite a while and afterward examine these surveys to find purchaser assessments on the perspectives through a feeling classifier lastly rank the item founded on significance of viewpoint by considering perspective recurrence and customers' perspectives given to every perspective over their general conclusions.

Surveys can be posted on the networks in three unique sorts:

- Type (1) - Pros and Cons: The analyst is approached to depict Pros and Cons independently.
- Type (2) - Pros, Cons and nitty gritty audit: The analyst is approached to depict Pros and Cons independently and furthermore compose a definite survey.
- Type (3) - free arrangement: The commentator can compose openly, i.e., no partition of Pros and Cons.

Various sorts of audits might require various strategies to play out the undertakings, for example, item perspective distinguishing proof, item opinion grouping and item viewpoint positioning.

To get distinguishing proof of perspectives, the Pros and Cons surveys are utilized as supporting information to help the ID of viewpoints in the free text audits. Specifically, first split the free message audits into sentences, and parse each sentence utilizing parser. After that the regular thing phrases are removed from the sentence parsing trees as up-and-comer perspectives.

Since these competitor viewpoints might contain clamors, further the Pros and Cons audits are utilized to help them in distinguishing proof of angles from the applicants. Then, at that point, all the regular thing terms extricated from the Pros and Cons surveys are gathered to frame a jargon.

Every perspective in the Pros and Cons audits is addressed into a unigram element, and every one of the viewpoints are then used to get familiar with a one-class Support Vector Machine (SVM) classifier. The resultant classifier is utilized to distinguish viewpoints in the up-and-comers extricated from the free text audits.

Maximum Entropy Classifier (ME)

Another classifier is Maximum Entropy classifier. The name Maximum Entropy comes from the way that the classifier observes the probabilistic model which is the easiest and least compelled. However it has a few explicit imperatives. The thought behind greatest entropy is that one ought to lean toward the most uniform models that additionally fulfill any given imperatives.

This Classifier is utilized to changes over marked highlight sets to vectors utilizing encoding. This encoded vector is then used to process loads for each component that can then be consolidated to decide the most probable name for a list of capabilities. This classifier is constrained by a bunch of X{weights}, which is utilized to consolidate the joint elements that are delivered from a list of capabilities by a X{encoding}. Each C {(feature set, label)} pair is planned to vector with encoding plan.

Support Vector Machines Classifiers (SVM)

Support Vector Machines (SVMs) are the most up to date directed AI method. SVM utilizes the thought of a "edge"- a hyperplane that partition two information classes.. An upper bound on the normal speculation mistake can be decreased by amplifying the edge and in this way biggest conceivable distance between isolating hyperplane cases on one or the other side of it. In, X, O are 2 classes and A, B and C are the three hyperplanes.

Hyperplane A gives the best partition between the classes, in light of the fact that the typical distance of any of the information focuses is the biggest, so it addresses the greatest edge of detachment. On account of directly divisible information, when the ideal isolating hyperplane is found, information focuses that lie on its edge are known as help vector focuses whose arrangement is addressed as a straight mix of just these places. Different information focuses are disregarded.

Thusly, the model intricacy of a SVM is unaffected by the quantity of highlights experienced in the preparation information. Therefore, SVMs are appropriate for learning errands where the quantity of elements is enormous as for the quantity of preparing examples.

DESCRIPTION OF THE INVENTION

Supervised learning

Administered learning procedure utilize the assortment of marked audits to get familiar with an extraction model. This extraction model called as extractor is then utilized for the recognizable proof of perspectives in ne audits. A large portion of the directed learning methods depend on the consecutive learning. Different literary works show the different method for the learning of extractor.

BRIEF DESCRIPTION OF THE DIAGRAM

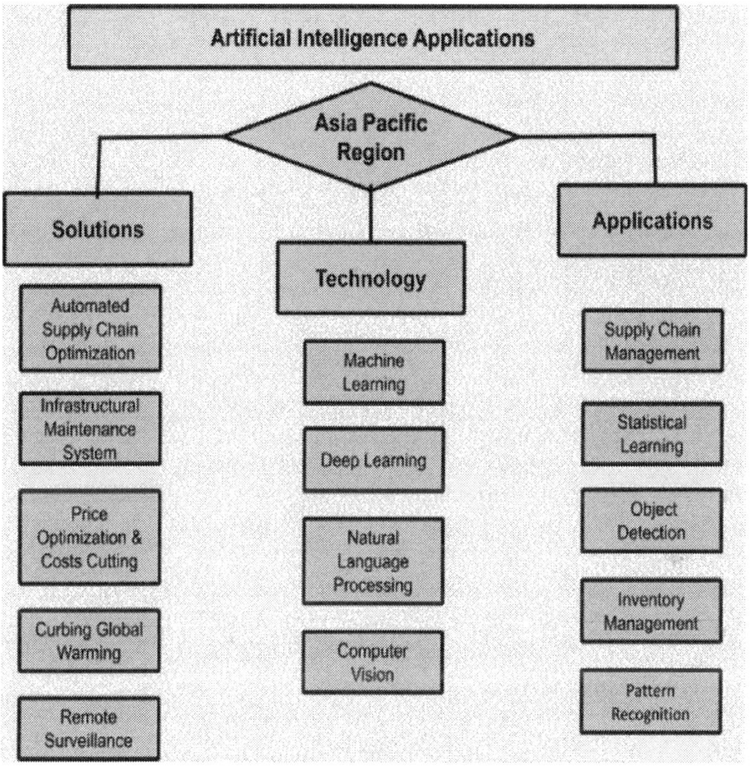

Fig. 1: AI-Based Product Ranking Evaluation Process System, Flow Chart.

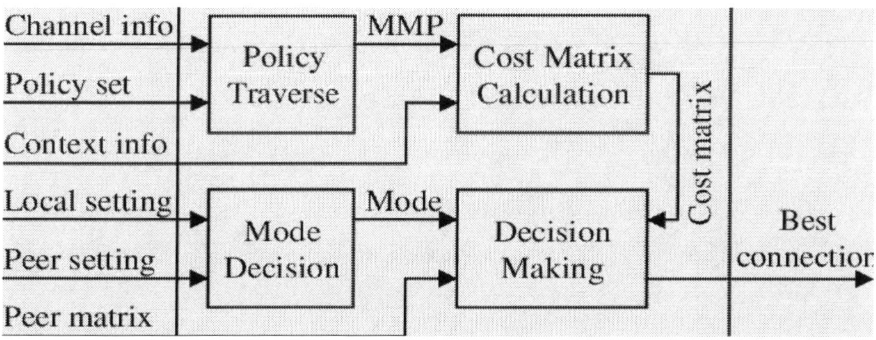

Fig. 2: AI- Based Product Ranking Evaluation Process System, Block Diagram.

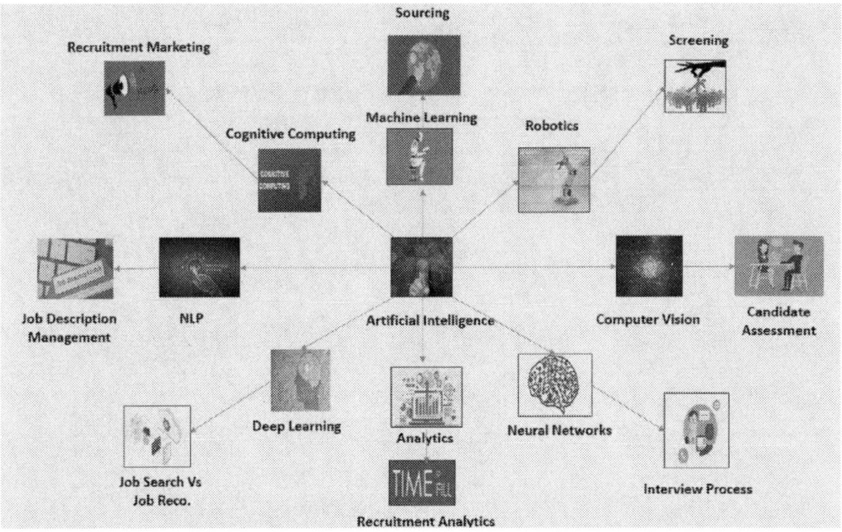

Fig. 3: AI-Based Product Ranking Evaluation Process System.

Wong and Lam utilizes the HMM model and restrictive irregular field to gain proficiency with the extractor.

Li et al utilizes skip CRF and tree CRF for example to incorporated CRF variety to become familiar with the extractor. The primary weakness of this technique is that it require Labeled example for preparing. These strategies are exceptionally tedious to mark the examples.

Unaided Learning

In this technique the perspectives are viewed as thing or thing stages and event recurrence of thing and thing phrases is determined. The regular thing or thing phrases are considered as perspectives. Hu and Liu utilize this solo method for perspective recognizable proof. Primary burden of this technique is that distinguished viewpoints applicants might contain clamor.

Wu et al utilizes an expression reliance parsing. Express reliance parsing accepts the sentence as information and sections it into phrases.

Then these sections are connected with coordinated bend. Express reliance parsing centers around phrases and not on single word inside state. To ensure that distinguished angles up-and-comer is to be viewpoint language model which depends on item surveys is utilized to anticipate score of up-and-comers. Model channel out low score competitors. Such model might be one-sided to visit terms in the audit and can't detect definitively the connected score of angle terms thus can't sift through clamor proficiently.

OUTCOME

Our Invention "Product Ranking using Machine Learning" is a An item might have many angles. A portion of the item angles are a higher priority than the others and have solid effect on the inevitable customer's decision making as well as association's item advancement systems. Recognizable proof of significant item perspectives become vital as the two buyers and firms are benefited by this. Buyers can without much of a stretch settle on buying choice by focusing on the significant viewpoints also as firms can zero in on working on the nature of these perspectives and in this manner upgrade item notoriety productively. This paper gives the depiction of different methods for item angle ID and grouping

- According to claim#1 the invention is to a "AI- Based Product Ranking Evaluation Process System" is a An item might have many angles. A portion of the item angles are a higher priority than the others and have solid effect on the inevitable customer's decision making as well as association's item advancement systems.
- According to claim#1,2 the invention is to a Recognizable proof of significant item perspectives become vital as the two buyers and firms are benefited by this. Buyers can without much of a stretch settle on buying choice by focusing on the significant viewpoints also as firms can zero in on working.
- According to claim#1,2,3 the invention is to a nature of these perspectives and in this manner upgrade item notoriety productively. This paper gives the depiction of different methods for item angle ID and grouping.

CITATIONS

- Dr. Divya Chirayil, Ms. Rupali Sathe, 202121047470, MAHATMA EDUCATION SOCIETY'S, PILLAI HOC COLLEGE OF ENGINEERING AND TECHNOLOGY," IoT-Kit to Track and Notify Doctor, Nurse and Medical Equipment" 2021, Indian Patent office.
- Gaurav Gyanendra Mishra, Akash Surendra Yadav, 202121046356, MAHATMA EDUCATION SOCIETY'S, PILLAI HOC COLLEGE OF ENGINEERING AND TECHNOLOGY, "ECO-FRIENDLY POLYPROPYLENE PAVER BLOCK WITH SWAPPABLE ANTI SKID FILM, 2021, Indian Patent office.
- Dr. Nivedita Shreyans (Associate Professor), 202121046795, MAHATMA EDUCATION SOCIETY'S, PIMSR-PILLAI INSTITUTE OF MANAGEMENT STUDIES AND RESEARCH, "Method and Process MCQ Examination Management System, 2021, Indian Patent office.
- Dr. Bhavna Dave (Assistant Professor), Mr. Hardik Dave (Assistant Professor), 202121048707, MAHATMA EDUCATION SOCIETY'S," English Speaking Error Detector AI- Based Mobile Chip, 2021, Indian Patent office.
- Dr. Sally Enos, Rohan Enoss Mathew, 202121048708, MES' PILLAI COLLEGE OF EDUCATION AND RESEARCH, Plot No 1, Sector-8, Khanda Colony, New Panvel (W), " Portable Tool to hold the Umbrella Travelling in a Car.", 2021, Indian Patent office.
- Dr. Reni Francis – Principal, Reuben Francis ,Ryan Francis, 351149-001, Design Patent, MAHATMA EDUCATION SOCIETY'S, MES' PILLAI COLLEGE OF EDUCATION AND RESEARCH- Chembur Mumbai. "Trio pen pencil and eraser," 2021, Indian Patent office.

Chapter - 03

Detection Faulty Three-Phase Induction Motor Using Machine Learning

Pragnesh Narendra Shah

Mahatma Education Society's, Pillai College of Engineering,
Dr. K. M. Vasudevan PILLAI Campus, Sector 16, New Panvel, NAVI Mumbai -410206, MH, India.

Prof (Dr.) B.K. Sarkar

MES's, Pillai HOC College of Engineering &Technology,
Rasayani Taluka Panvel, Dist, Navi Mumbai, Maharashtra 410207, India.

ABSTRACT

Our Invention "Detection Faulty Three-Phase Induction Motor using Machine Learning" is a methodology for ID of stator winding open circuit issue in enlistment engine in light of the d-hub versus q-hub current plot. Inception of open circuit in one of the stator windings causes uneven winding condition, which powers the engine to draw lopsided current from the inventory. This unbalance in stator current reflects in dq0 changed stator current parts and is utilized to identify the advancement of open circuit issue. The d-hub versus q-hub current plot is round in nature during sound state of the engine and at broken condition, the plot reshaped to a circle. The tendency of significant hub of oval concerning x-hub changes with shortcoming seriousness. A MATLAB/Simulink model of a line took care of three-stage acceptance engine is created to concentrate on the engine conduct during shortcoming. Reproduction model of enlistment engine for open circuit issue and the shortcoming location procedure revalidated by trial brings about constant and also the three-stage acceptance engines are viewed as one of the main components of the modern cycle. Be that as it may, in this climate, these machines are dependent upon electrical and mechanical issues, which might cause critical monetary misfortunes. Consequently, the reason for this paper is to introduce an example acknowledgment strategy for the identification of stator windings short-circuits in view of proportions of shared data between the stages current signs.

FIELD

Our invention is related to a Detection of stator incipient faults and identification of faulty phase in three-phase induction motor and also the proposed designs, highlight vectors acquired from ordinary and defective engines are applied to two geographies of counterfeit brain organizations. The arrangement results introduced exactness's more than 93% in any event, when the engines were dependent upon a few states of burden force and power supply voltage unbalance.

The aim of issue location is to distinguish the issue toward the early phase and turned down the machine promptly to stay away from engine disappointment because of the enormous issue current. In this work, an internet based shortcoming finding of stator interterm issue of a three-stage acceptance engine in light of the idea of balanced parts is introduced.

BACKGROUND OF THE INVENTION

Among the electrical disappointments, up to 36% of the cases are expected deceased benefactor winding shortcomings. Explicitly with respect to this sort of issue Discrete Wavelet Transform (DWT) is utilized to get designs from high recurrence parts that are characterized utilizing Artificial Neural Networks (ANN).

Likewise, factual proportions of the stator current, like mean, difference, skewness, and kurtosis, are used as contributions to a Multi-layer perceptron (MLP) ANN. Additionally, in these actions are likewise utilized for rotor capriciousness flaws alongside stator winding between turn impede

Customary methods of shortcoming observing were with respect to the spillage transition detecting halfway current and voltage and so forth Resulting studies, in any case, exhibited that large numbers of these customary strategies are [4] obligated to peculiarity because of supply voltage mutilations worked in machine deviations incidental impacts of the stator and rotor issues.

The forward [5] motor current signature analysis (MCSA) is a huge technique embraced for condition checking. Acknowledgment motor [2,3] issues like bearing issues, broken rotor bar, flightiness abnormalities, and stator winding imperfections cause assortment in abundancy and repeat of motor [4] current imprint.

A jump forward in signal embellishment strategies and advances in PC programming has taken machine deficiency disclosure to fresher heights. By far most of the displayed work in stator winding [4,5] issue acknowledgment is from a repeat examination space. Signal changing procedures like Fast Fourier Transform (FFT), Transform, short [4,3] period of time Fourier Transform (STFT), wavelet Transform, and Hilbert Transform have been taken on in blend in with [2,3,4] various request techniques like expert systems, counterfeit cerebrum association, soft reasoning, and support vector machine for the motor degradation.

Unmistakable interest has been shown in for counterfeit mind network in weakness revelation of the acknowledgment motor. [3,2] The significant condition to creation a productive ANN classifier is the assurance of appropriate commitments with respect to each case of deficiency. [1,2,3] In stator interterm deficiency region acknowledgment by ANN is portrayed contemplating repeat space limits as picked input?

In ANN is applied for recognizing [4] the earnestness of the intercoil deficiency with the picked limit from time region. The work in spotlights on awful [3,5] gathering current which is caused in view of lopsided windings.

PRIOR ART SEARCH

https://www.researchgate.net/publication/281309486_Detection_of_stator_incipient_faults_and_identification_of_faulty_phase_in_three-phase_induction_motor_-_simulation_and_experimental_verification

https://digital-library.theiet.org/content/journals/10.1049/iet-epa.2015.0024?crawler=redirect

https://ieeexplore.ieee.org/document/8707681

Chrome-extension://efaidnbmnnnibpcajpcglclefindmkaj/viewer. html?pdfurl=https%3A%2F%2Fwww.ijrte.org%2Fwp-content%2Fuploads%2Fpapers%2Fv4i4%2FD1475094415.pdf&clen=974645&chunk=true

Chrome-extension:// efaidnbmnnnibpcajpcglclefindmkaj/ viewer. html? pdfurl=https%3A%2F%2Fisiarticles.com%2Fbundles%2FArticle%2Fpre%2Fpdf%2F144153.pdf&clen=232467&chunk=true

OBJECTIVES OF THE INVENTION

- The objective of the invention is to provide a "Detection Faulty Three-Phase Induction Motor using Machine Learning" is a methodology for ID of stator winding open circuit issue in enlistment engine in light of the d-hub versus q-hub current plot. Inception of open circuit in one of the stator windings causes uneven winding condition, which powers the engine to draw lopsided current from the inventory.
- The other objective of the invention is to provide a unbalance in stator current reflects in dq0 changed stator current parts and is utilized to identify the advancement of open circuit issue. The d-hub versus q-hub current plot is round in nature during sound state of the engine and at broken condition, the plot reshaped to a circle. The tendency of significant hub of oval concerning x-hub changes with shortcoming seriousness.
- The other objective of the invention is to provide a MATLAB/ Simulink model of a line took care of three-stage acceptance engine is created to concentrate on the engine conduct during shortcoming. Reproduction model of enlistment engine for open circuit issue and the shortcoming location procedure revalidated by trial brings about constant and also the three-stage acceptance engines are viewed as one of the main components of the modern cycle.
- The other objective of the invention is to provide a these machines are dependent upon electrical and mechanical issues, which might cause critical monetary misfortunes. Consequently, the reason for this paper is to introduce an example acknowledgment strategy for the identification of stator windings short-circuits.

SUMMARY OF THE INVENTION

The most widely recognized abandons related with TIM stator are phase to ground, stage to stage, and curl to loop short-circuits. When there is a short out including only a couple of loop turns, the stator insulation begins to fall apart; this could quickly form into love genuine issue and influence the engine activity.

In request to screen the TIM working circumstances and identify these flaws, one should utilize an appropriate sign handling technique for the extraction and choice of applicable elements of the signals. The engine current mark investigation is a normally utilized method because it is basic, painless, and can be robotized. This approach is centered on the current otherworldly investigations and has been widely utilized for TIM shortcoming analysis.

Mathematical Modeling

[1] A three-stage enrolment motor with turn issue in one-stage stator winding is considered, where [1,2] β is the little part of shorted turns. Contorting in this stage has two segments shorted turns and fearless turns.

[4,5] Machine conditions in abc factors for a reasonable motor with turn weakness in one winding can be imparted Here, we [4,6] acknowledged that the spillage inductance of the shorted turns is, the place where is the per-stage spillage inductance and the issue impedance is resistive.

Arrangement Component Analysis

The even parts are strong device for dissecting and tackling the issues of any uneven framework. The balanced parts are observed dependable marks of stator turn issues. By standard, balanced (without shortcoming) engines provided with even three-stage voltage sources won't deliver negative grouping flows. At the point when a turn issue happens, evenness will upset and create negative and zero succession flows.

Regarding the even parts practice, three arrangements of even adjusted stages are gotten from any arrangement of uneven boundaries. They are perceived as certain, negative, and zero arrangement parts.

BRIEF DESCRIPTION OF THE DIAGRAM

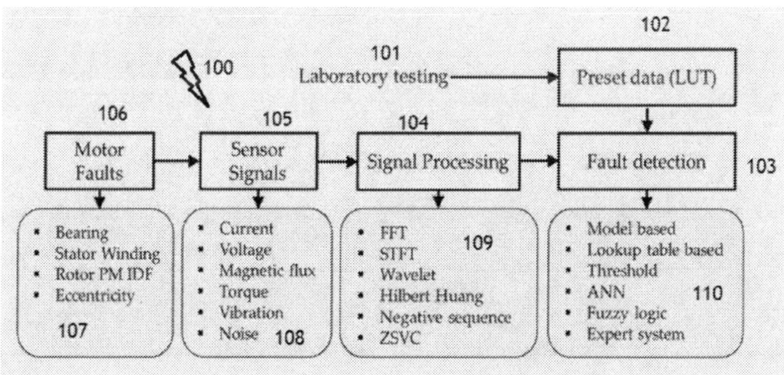

Fig. 1: Detection of stator incipient faults and identification of faulty phase in three-phase induction motor Flow chart.

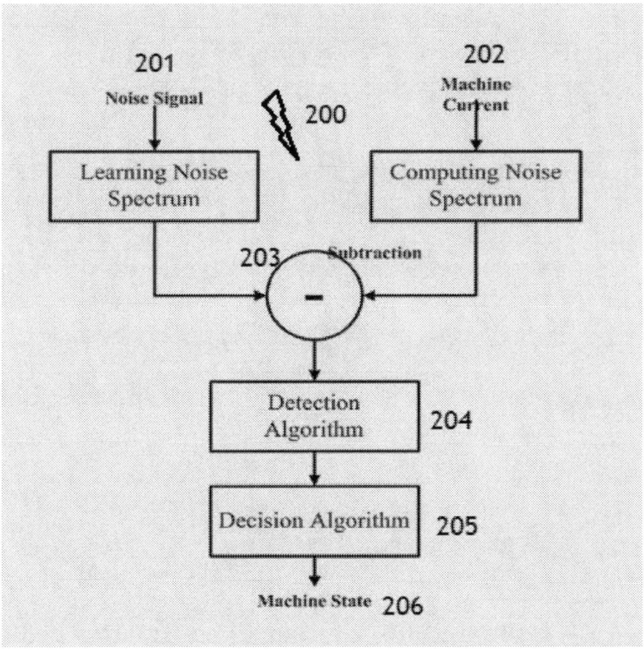

Fig. 2: Detection of stator incipient faults and identification of faulty phase in three-phase induction motor, Block Diagram.

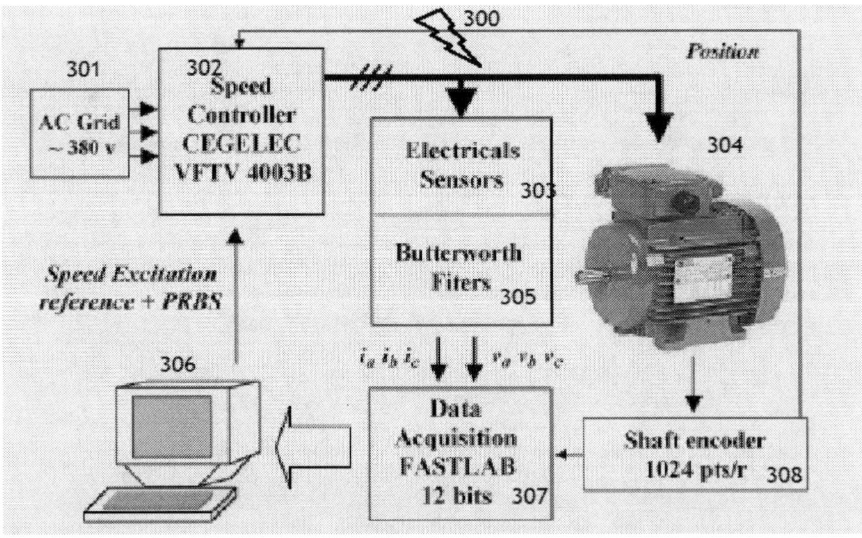

Fig. 3: Detection of stator incipient faults and identification of faulty phase in three-phase induction motor.

DESCRIPTION OF THE INVENTION

MATLAB Ssoftware

[5] Is used for making a generation model of a three-stage motor with turn deficiency in one of the stage winding. In light of the difficulty in [3,4] making the issue and assessing the stage streams likely for high potential gains of rate shorting, we are constrained to make the Simulink model. Simulink of the selection motor with stator winding shorting is [4] created considering the focal. The Simulink model of the motor with interterm short.

Network for Classification

[1] Counterfeit brain networks are liberal to commotion and answer rapidly, so they can be utilized continuously shortcoming location. Since it is beyond the realm of possibilities to expect to make a look-into table [3]putting away information for all circumstances, a feedforward brain network is utilized for characterizing the shortcoming.

Expecting most [3,4] prominent precision from the pre-arranged mind association, input vector is made including possible exploratory characteristics and Simulink values for high rate shorting. [4,5]The various cycles related with created by area of earnestness level of shorting in stator winding

The display of the computation is tricky to the setting of the learning rate. A minuscule learning rate will provoke [3,4] longer joining time, and an extraordinarily high learning rate can incite influencing and unstable estimation.

Backpropagation getting ready with a flexible learning rate is executed with point dive work. Overstated diversion sigmoid trade work is used, which discovers a layer's outcome from its net data. [2,3,4] Mean squared normalized screw up execution work gauges the association execution as demonstrated by the mean of squared botches, when [3,4,5] joined into the planning communication redesigns the adequacy of the synaptic weight's change. [2,3,4] A very low MSE mirrors that the ideal outcome and the ANNs yields are almost each other, and thusly the association is completely ready.

Result and Performance Validation

While dissecting the presentation and relapse plot of organizations, it has been observed that brain networks responded well with preparing and approval tests. [3,4,5] Approval execution of the organization gave

100 percent exactness (with 50 examples).

Along these lines, precision rate is determined with blunder between target esteem and real result vector. [3,4,5] The degree of precision got is 99.75%. This examination center is to gain ground as well as to work on the field of condition observing and shortcoming location in an enlistment engine. [1,2,3] Negative arrangement current observing is one of the most straightforward yet strong and significant procedures for stator impede.

It is applicable to determine that determination of shortcoming marker is extremely basic in the characterization cycle. [3,4] In this work, per unit change in a negative grouping current with positive arrangement current is considered as the issue marker thus observed [1] to be more summed up method for interterm stator winding issue identification.

While stage current is considered as shortcoming marker. [2] When taken δ as shortcoming marker, input vector of [4] NN is no different for various engines in view of its per unit nature.

The issue boundary viable is gotten from the time space, which dodges the utilization of complex sign molding procedures utilized in recurrence area for issue identification.

The work presents a [5] utilization of brain organization to order the stator interterm issue. The organization is prepared [2,3] with a full scope of info vector utilizing trial values [1,4] (for a little degree of shortcoming) as well as Simulink values (for significant degree of issue).

Hence, [2,4] NN is thoroughly prepared with a full scope of informational indexes. Simulink displaying makes a boundless data set, which is preposterous through test. [3,5,7] The exhibition of the NN is viewed as precise and quick. Shortcoming discovery in the early phase expands the plausibility of [2,4]fixing the machine and maintains a strategic distance from the gamble of fire and blast.

OUTCOME

Our Invention "Detection of stator incipient faults and identification of faulty phase in three-phase induction motor" is a methodology for ID of stator winding open circuit issue in enlistment engine in light of the d-hub versus q-hub current plot. Inception of open circuit in one of the stator windings causes uneven winding condition, which powers the engine to draw lopsided current from the inventory. This unbalance in stator current reflects in dq0 changed stator current parts and is utilized to identify the advancement of open circuit issue. The d-hub versus q-hub current plot is round in nature during sound state of the engine and at broken condition, the plot reshaped to a circle. The tendency of significant hub of oval concerning x-hub changes with shortcoming seriousness. A MATLAB/Simulink model of a line took care of three-stage acceptance engine is created to concentrate on the engine conduct during shortcoming. Reproduction model of enlistment engine for open circuit issue and the shortcoming location procedure revalidated by trial brings about constant and also the three-stage acceptance engines are viewed as one of the main components of the modern cycle. Be that as it may, in this climate, these machines are dependent upon electrical and mechanical issues, which might cause critical monetary misfortunes. Consequently, the reason for this paper is to introduce an example acknowledgment strategy for the identification of stator windings short-circuits in view of proportions of shared data between the stages current signs. According to claim1# the invention is a "Detection of stator incipient faults and identification of faulty phase in three-phase induction motor" is a methodology for ID of stator winding open circuit issue in enlistment engine in light of the d-hub versus q-hub current plot. Inception of open circuit in one of the stator windings causes uneven winding condition, which powers the engine to draw lopsided current from the inventory.

- According to claim1,2# the invention is a unbalance in stator current reflects in dq0 changed stator current parts and is utilized to identify the advancement of open circuit issue. The d-hub versus q-hub current plot is round in nature during sound state of the engine and at broken condition, the plot reshaped to a circle. The tendency of significant hub of oval concerning x-hub changes with shortcoming seriousness.
- According to claim1,2,3# the invention is a MATLAB/Simulink model of a line took care of three-stage acceptance engine is created to concentrate on the engine conduct during shortcoming. Reproduction model of enlistment engine for open circuit issue and the shortcoming location procedure revalidated by trial brings about constant and also the three-stage acceptance engines are viewed as one of the main components of the modern cycle.
- According to claim1,2,3,4# the invention is a these machines are dependent upon electrical and mechanical issues, which might cause critical monetary misfortunes. Consequently, the reason for this paper is to introduce an example acknowledgment strategy for the identification of stator windings short-circuits in view of proportions of shared data between the stages current signs.

CITATIONS

- Dr. Reni Francis – Principal, Reuben Francis ,Ryan Francis 351150-001, Design Patent , MAHATMA EDUCATION SOCIETY'S, MES' PILLAI COLLEGE OF EDUCATION AND RESEARCH- Chembur Mumbai., "All in One Geometry Set," 2021, Indian Patent office.
- MAHATMA EDUCATION SOCIETY 'S, 511510-001, MES's Pillai College of Education and Research, Sector-8, Khanda Colony, New Panvel East, Navi Mumbai, Maharashtra 410207, India., "MES SPORK" 2021, Indian Patent Office.
- Dr. Lata Menon, Mr. Ravi Bari, Ms. Neethumol K G, 202121050104, MES's PILLAI HOC COLLEGE OF ENGINEERING &TECHNOLOGY, RASAYANI, "Bitcoin Price Quick Prediction Using AI, DL and ML", 2021, Indian Patent Office.
- Dr. Divya Chirayil, Ms. Rupali Sathe, Abhishek Jain, 202141050116, MES's PILLAI HOC COLLEGE OF ENGINEERING &TECHNOLOGY, RASAYANI, "Emotion Recognition and notification using Machine Learning and AI" 2021, Indian Patent Office.
- Dr. B.K.Sarkar, 20211105174, MAHATMA EDUCATION SOCIETY'S ,PILLAI HOC COLLEGE OF ENGINEERING AND TECHNOLOGY, "Advanced Branding in the Marketing of Unilever Products using ML" 2021, Indian Patent office.
- Dr. Sandeep S. Shinde, Assist. Professor, Dr. Sejal Rohit Tambake, 202121052143, (O)MAHATMA EDUCATION SOCIETY, S VIDYADHIRAJA COLLEGE OF PHYSICAL EDUCATION AND RESEARCH, "Break in Chant detector of a Raider Playing Kabaddi" 2021, Indian Patent office.
- Dr. Mansi Subhedar (Associate Professor), Mr. Upendra Patil (Assistant Professor),202121052142, MAHATMA EDUCATION SOCIETY'S, PILLAI HOC COLLEGE OF ENGINEERING AND TECHNOLOGY," Light Weight Transform Domain IMAGE Steganography Using 32-bit ARM Cortex-M4 "2021, Indian Patent office.
-

Chapter - 04

Big Data Regression Analysis Using Machine Learning

Khedekar Vilas Baburao

Research Scholar, School of Computer Science and Engineering, Vellore Institute of Technology, Vellore, India.
Assistant Professor, MIT Art Design and Technology University, Pune

Dr. Dharmendra Singh Rajput

School of Information Technology & Engineering
Associate Professor, Vellore Institute of Technology, Vellore, India

Reuben Francis

M. Tech SY Student, Mit ADT School of Engineering, Pune

ABSTRACT

Our Invention "Point Regression Analysis-Based Big Data Processing Using Machine Learning and Deep Learning" is a In the current age of the Fourth Industrial Revolution (4IR or Industry 4.0), the computerized world has an abundance of information, like Internet of Things (IoT) information, network protection information, versatile information, business information, web-based entertainment information, wellbeing information, and so on To brilliantly investigate these information and foster the relating shrewd and mechanized applications, the information on man-made consciousness (AI), especially, AI (ML) is the key. Different sorts of AI calculations, for example, directed, solo, semi-managed, and support learning exist nearby. Moreover, the profound realizing, which is essential for a more extensive group of AI strategies, can astutely examine the information for an enormous scope. In this paper, we present an extensive view on these AI calculations that can be applied to upgrade the knowledge and the abilities of an application. Consequently, this study's key commitment is making sense of the standards of various AI procedures and their pertinence in different true application spaces, for example, network safety frameworks, shrewd urban areas, medical care, web based business, farming, and some more. We additionally feature the difficulties and potential examination headings in light of our review. In general, this paper plans to fill in as a kind of perspective point for both scholarly community and industry experts as well with respect to chiefs in different certifiable circumstances and application regions, especially according to the specialized perspective.

Key: *Big Data, Regression, Analysis, Machine Learning, procedures, AI calculations, versatile, business information.*

BACKGROUND

Separating experiences from these information can be utilized to assemble different canny applications in the significant areas. For example, to fabricate an information driven robotized and shrewd network protection framework, the applicable network protection information can be utilized. to assemble customized setting mindful savvy portable applications, the pertinent versatile information can be utilized, etc.

Hence, the information the executives devices and procedures having the capacity of removing experiences or valuable information from the information in an ideal and smart manner is critically required, on which this present reality applications are based.

The ubiquity of these ways to deal with learning is expanding step by step, which is displayed light of information gathered from Google Trends in the course of the most recent five years. The x-hub of the figure demonstrates the particular dates and the relating prominence score inside the scope of 0(minimum) 0(minimum) to 100(maximum) 100 (greatest) has been displayed in y-pivot.

Agreeing the prevalence sign qualities for these learning types are low in 2015 and are expanding step by step. These insights spur us to study on AI in this paper, which can assume a significant part in reality through Industry 4.0 robotization.

OBJECTIVES

To characterize the extent of our review by considering the nature and qualities of different kinds of genuine information and the capacities of different learning methods.

- To give an exhaustive view on AI calculations that can be applied to upgrade the insight and capacities of an information driven application.

- To examine the immaterialness of AI based arrangements in different genuine application areas.
- To feature and sum up the potential examination headings inside the extent of our review for astute information investigation and administrations.

SUMMARY

In light of the significance and possibility of "AI" to dissect the information referenced above, in this paper, we give a thorough view on different kinds of AI calculations that can be applied to improve the insight and the abilities of an application.

Consequently, the vital commitment of this study is making sense of the standards and possibility of various AI procedures, and their relevance in different true application regions referenced before.

The motivation behind this paper is, consequently, to give an essential manual for those scholarly community and industry individuals who need to study, research, and foster information driven mechanized and smart frameworks in the applicable regions in light of AI procedures.

Classification Analysis

Arrangement is viewed as a managed learning technique in AI, alluding to an issue of prescient demonstrating too, where a class mark is anticipated for a given model. Numerically, it maps a capacity (f) from input factors (X) to yield factors (Y) as target, name or classes.

To anticipate the class of given informative items, it very well may be done on organized or unstructured information. For instance, spam discovery, for example, "spam" and "not spam" in email specialist co-ops can be an order issue. In the accompanying, we sum up the normal order issues.

Parallel Order

It alludes to the arrangement assignments having two class marks, for example, "valid and bogus" or "yes and negative".

In such twofold grouping undertakings, one class could be the typical state, while the unusual state could be another class. For example, "disease not recognized" is the typical condition of an undertaking that includes a clinical trial, and "malignant growth distinguished" could be considered as the unusual state. Essentially, "spam" and "not spam" in the above illustration of email specialist organizations are considered as twofold characterization.

Multiclass Classification

Generally, this alludes to those characterization undertakings having in excess of two class marks the multiclass order doesn't have the guideline of typical and strange results, dissimilar to paired grouping assignments. All things being equal, inside a scope of determined classes, models are delegated having a place with one.

For instance, it very well may be a multiclass characterization errand to order different sorts of organization assaults in the NSL-KDD dataset, where the assault classifications are grouped into four class names, like DoS (Denial of Service Attack), U2R (User to Root Attack), R2L (Root to Local Attack), and Probing Attack.

Multi-mark Characterization

In AI, multi-mark characterization is a significant thought where a model is related with a few classes or names. Accordingly, it is a speculation of multiclass grouping, where the classes engaged with the issue are progressively organized, and every model may at the same time have a place with more than one class in each various leveled level, e.g., staggered text characterization.

For example, Google news can be introduced under the classes of a "city name", "innovation", or "most recent news", and so forth Multi-mark arrangement incorporates progressed AI calculations that help anticipating different commonly non-elite classes.

DIAGRAM

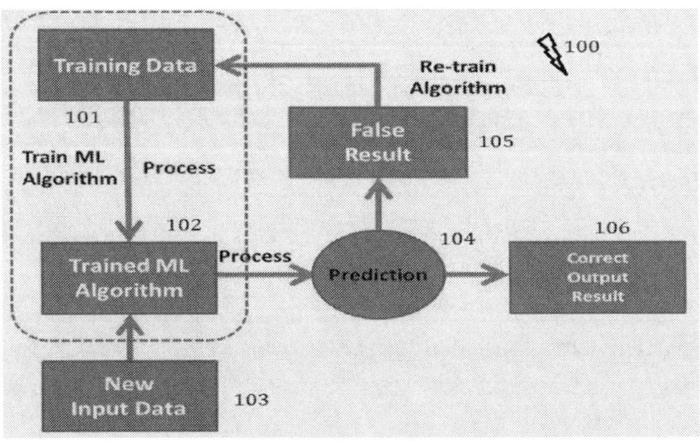

Fig.1: Point Regression Analysis-Based Big Data Processing Using Machine Learning and Deep Learning, Flow Chart.

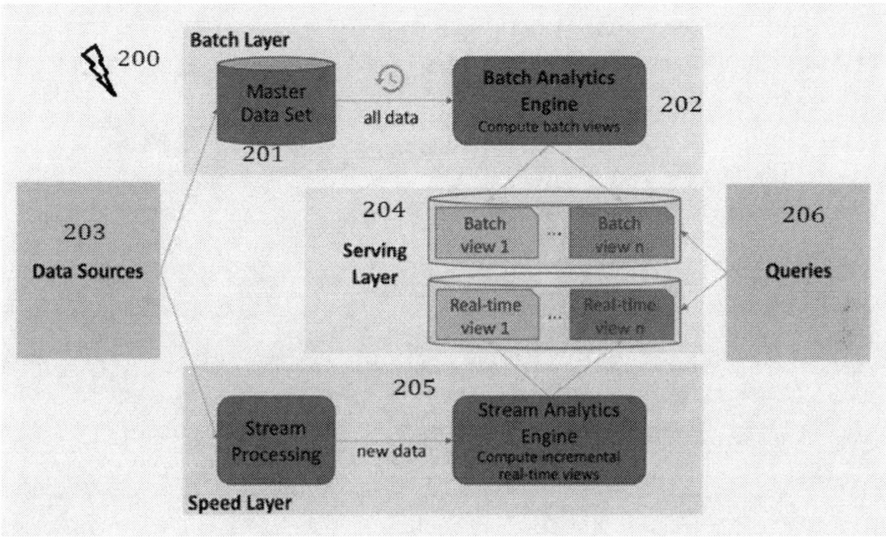

Fig. 2: Point Regression Analysis-Based Big Data Processing, Block Diagram

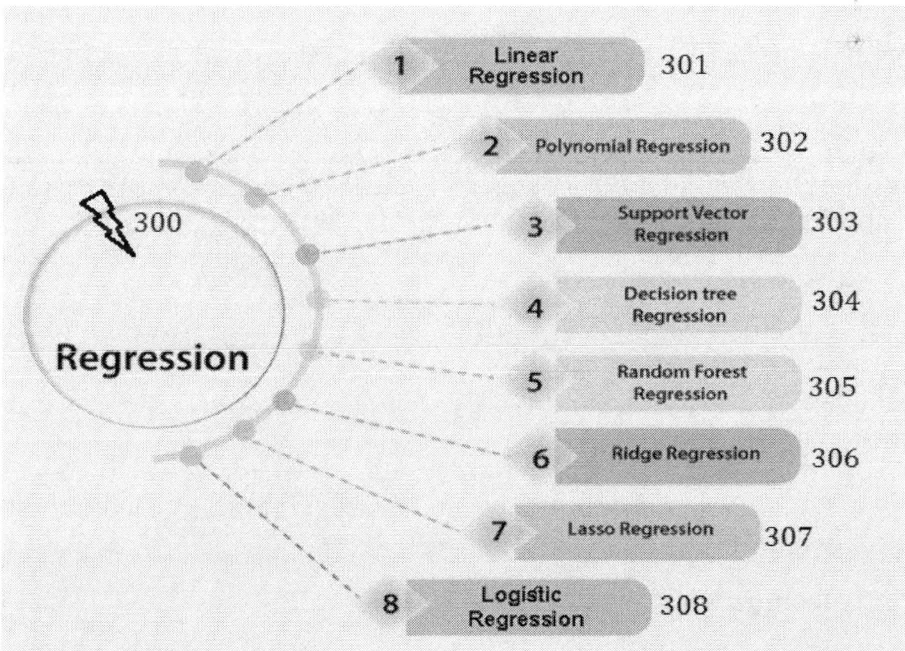

Fig. 3: Point Regression Analysis-Based Big Data Processing.

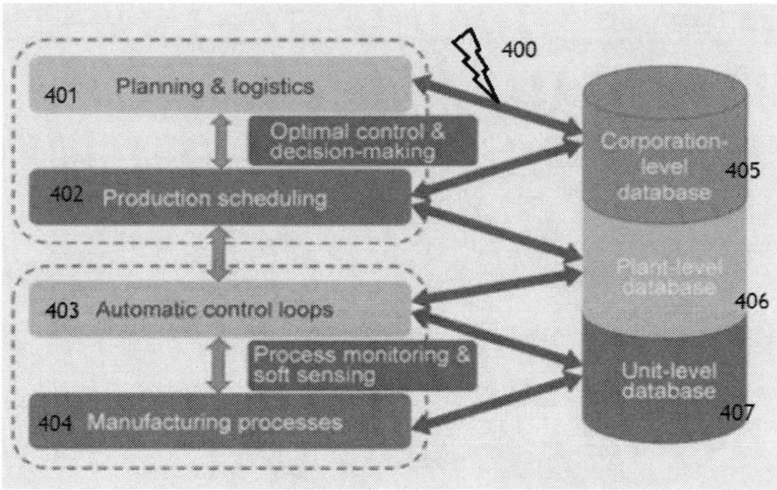

Fig. 4: Analysis-Based Big Data Processing Using Machine Learning and Deep Learning,

DESCRIPTION OF THE INVENTION

Generally, the accessibility of information is considered as the way to develop an AI model or information driven genuine frameworks Data can be of different structures, for example, organized, semi-organized, or unstructured. Plus, the "metadata" is another sort that ordinarily addresses information about the information. In the accompanying, we momentarily talk about these sorts of information.

Organized

It has an obvious construction, adjusts to an information model observing a guideline request, which is exceptionally coordinated and handily got to, and utilized by a substance or a PC program. In obvious plans, for example, social data sets, organized information are regularly put away, i.e., in an even organization. For example, names, dates, addresses, charge card numbers, stock data, geolocation, and so on are instances of organized information.

Unstructured

Then again, there is no pre-characterized configuration or association for unstructured information, making it considerably more challenging to catch, process, and investigate, generally containing text and mixed media material. For instance, sensor information, messages, blog passages, wikis, and word handling archives, PDF records, sound records, recordings, pictures, introductions, website pages, and numerous different sorts of business reports can be considered as unstructured information.

Semi-structured

Semi-organized information are not put away in a social data set like the organized information referenced above, yet it has specific hierarchical properties that make it more straightforward to investigate. HTML, XML, JSON records, NoSQL data sets, and so forth, are a few instances of semi-organized information.

Metadata

It isn't the typical type of information, yet "information about information". The essential distinction among "information" and "metadata" is that information are just the material that can characterize, measure, or even record something comparative with an association's information properties.

Then again, metadata portrays the important information data, giving it more importance for information clients. An essential illustration of an archive's metadata may be the creator, record size, date produced by the report, watchwords to characterize the record, and so fort.

Research

Our Invention "Point Regression Analysis-Based Big Data Processing Using Machine Learning and Deep Learning" is a In the current age of the Fourth Industrial Revolution (4IR or Industry 4.0), the computerized world has an abundance of information, like Internet of Things (IoT) information, network protection information, versatile information, business information, web-based entertainment information, wellbeing information, and so on To brilliantly investigate these information and foster the relating shrewd and mechanized applications, the information on man-made consciousness (AI), especially, AI (ML) is the key. Different sorts of AI calculations, for example, directed, solo, semi-managed, and support learning exist nearby. Moreover, the profound realizing, which is essential for a more extensive group of AI strategies, can astutely examine the information for an enormous scope. In this paper, we present an extensive view on these AI calculations that can be applied to upgrade the knowledge and the abilities of an application. Consequently, this study's key commitment is making sense of the standards of various AI procedures and their pertinence in different true application spaces, for example, network safety frameworks, shrewd urban areas, medical care, web based business, farming, and some more. We additionally feature the difficulties and potential examination headings in light of our review. In general, this paper plans to fill in as a kind of perspective point for both scholarly community and industry experts as well with respect to chiefs in different certifiable circumstances and application regions, especially according to the specialized perspective.

- According to claim1# the invention is to a "Point Regression Analysis-Based Big Data Processing Using Machine Learning and Deep Learning" is a in the current age of the Fourth Industrial Revolution (4IR or Industry 4.0), the computerized world has an abundance of information.
- According to claim1,2# the invention is to a IoT information, network protection information, versatile information, business information, web-based entertainment information, wellbeing information, and so on To brilliantly investigate these information and foster the relating shrewd and mechanized applications, the information on man-made consciousness (AI), especially, AI (ML) is the key.
- According to claim1,2,3# the invention is to a "Different sorts of AI calculations, for example, directed, solo, semi-managed, and support learning exist nearby. Moreover, the profound realizing, which is essential for a more extensive group of AI strategies, can astutely examine the information for an enormous scope. In this paper, we present an extensive view on these AI calculations that can be applied to upgrade the knowledge and the abilities of an application.
- According to claim1,2,3,4# the invention is to a "Consequently, this study's key commitment is making sense of the standards of various AI procedures and their pertinence in different true application spaces, for example, network safety frameworks, shrewd urban areas, medical care, web based business, farming, and some more. We additionally feature the difficulties and potential examination headings in light of our review.
- According to claim1,2,3,4,5# the invention is to a "plans to fill in as a kind of perspective point for both scholarly community and industry experts as well with respect to chiefs in different certifiable circumstances and application regions, especially according to the specialized perspective.

CITATIONS

- Dr. B.K.Sarkar, 202111051548, MAHATMA EDUCATION SOCIETY'S, PILLAI HOC COLLEGE OF ENGINEERING AND TECHNOLOGY, "Home Based Digital FM Transmission using Machine Learning and IoT based System", 2021, Indian Patent office.
- Dr. Lata Menon, Dr. Archana Bhagwat, Dr. Vishakha Telgote, 202121050103," MAHATMA EDUCATION SOCIETY'S PILLAI HOC COLLEGE OF ARTS, SCIENCE AND COMMERCE RASAYANI", Isolation and Characterization of Bioactive Unique Complex Compounds from Stembark Advanced Extract of Uapaca Pilosahutch, 2021, Indian Patent office.
- Dr. Lata Menon, Dr. Jayanta Kumar Behera, Dr. Sapana Chilate,202141051543, MAHATMA EDUCATION SOCIETY'S PILLAI HOC COLLEGE OF ARTS, SCIENCE AND COMMERCE RASAYANI TALUKA KHALAPUR DISTRICT RAIGAD MAHARASHTRA 410207, INDIA., " Power Generation from Sea Wave and Create Renewable Energy" 2021, Indian Patent office.
- Dr. B.K.Sarkar, 20211050099, MAHATMA EDUCATION SOCIETY'S, PILLAI HOC COLLEGE OF ENGINEERING AND TECHNOLOGY, "Online Technical Learning and Advanced Training Management System" 2021, Indian Patent office.
- Mr. Rajendra Bhimrao Deshmukh; 202121050101, MAHATMA EDUCATION SOCIETY'S , PILLAI COLLEGE OF EDUCATION AND RESEARCH, " Intelligent Biometric Sanitizer with Academy Tools" 2021, Indian Patent office.
- Dr. Mansi Subhedar (Associate Professor), Mr. Jayesh Rane (Assistant Professor),202121052899,MAHATMA EDUCATION SOCIETY'S; PILLAI HOC COLLEGE OF ENGINEERING &TECHNOLOGY;" Black Box for Two Wheelers" 2021, Indian Patent office.
- Dr. Priam Pillai (Associate Professor), Amar Arun Jadhav ,(Assistant Professor)202121052898, MAHATMA EDUCATION SOCIETY'S, PILLAI COLLEGE OF ENGINEERING, PANVEL, "Antislip and Quick Release Grippers to Handle Soft and Delicate Materials" 2021, Indian Patent office.

CHAPTER - 05

Improved Performance of Passive Direct Methanol Fuel Cell Using Machine Learning

N V Raghavaiah

Research Scholar, Department of Mechanical Engineering, National Institute of Technology, Warangal, Email: nvraghavaiah@student.nitw.ac.in, nvraghavaiah@yahoo.co.in

I. Hari Prasad

Research Scholar, Dept of Mechanical Engineering, National Institute of Technology, Warangal, Telangana State, India. Email: ihprasad@gmail.com, ihprasad@student.nitw.ac.in

Dr. G. Naga Srinivasulu

Associate Professor, Dept of Mechanical Engineering, NIT Warangal, India. Email: gns@nitw.ac.in

ABSTRACT

Our Invention "Improved Performance of Passive Direct Methanol Fuel Cell using Machine Learning" is a creation connects with performance and fundamental perspectives concerning the Passive Direct Methanol Fuel Cell innovation that are mainly focusing on Membrane Electrode Assembly effective areas and perimeters. Hence it is required to analyze the best performance and to find out the highest power density of cell using various shapes by altering gasket openings of dynamic regions with equal areas and equal perimeters. Comparison of the shapes such as square, rectangular, rhombus, and circular active area forms and by using Nickel as current collector material is selected for evaluation. In the equal areas with gasket opening area of 1963.5 (+/-0.2) mm2, among circular, rhombus, rectangular and square shapes chosen for investigation on the performance of cell at 5 molar methanol concentration, square shape opening consisting of the perimeter of 177.2 mm has developed highest power density of 6.344 mWcm-2with a corresponding maximum current density of 65.2 mAcm-2. In the equal perimeters of 157.1 (+/- 0.2) mm of gasket openings, among circular, rhombus, rectangular and square shapes chosen for investigation on the performance of cell at 5 molar methanol concentration, Rhombus shape opening consisting of area 1400 mm2has developed the highest power density of 7.714 mWcm-2 with a corresponding maximum current density of 85.3 mAcm-2. In the large active areas, methanol consumption is slow because the current density is lower when compared to the smaller active areas. As a result, the surface-specific concentration on the ion-conducting membrane is being more, making a higher crossover of methanol over bigger active zones. The experimental investigation demonstrated that the size of the dynamic zone significantly affects the performance of the cell and its performance can be improved with the expansion in dynamic region size. In the vast dynamic regions, the current density is lower compared to the smaller dynamic zones where the methanol is expended gradually. So, the local concentration on the cell membrane is more and brings further fuel cross-over. The exothermic response that is involved on penetrated fuel and oxidant on cathode impetus generates more heat, which prompts an elevated working temperature. This elevated working temperature brings about superior electrochemical kinetics of fuel oxidation and a decrease in oxygen response, leading to superior cell performance. The internal obstruction decreases with expanding the size of the dynamic zone. This reality is good with the accomplished outcome. Even though the cell performance can be enhanced by expanding the size of the dynamic zone, with the research outcome as on date, it is revealed that tiny dynamic zones have its focal points, for example, more slow charge transfer obstruction, lesser flooding, even distribution of methanol dispersion, increasing of effective current gathering, minor expenditure on fabrication, and further, it is small in size and weight.

It is also learned from the experiment that cell performance depends on the combined effect of (i) gasket opening shape, opening area, and its perimeter; (ii) the free open ratio and opening perimeter on the current collector plates. Comparison study referring to power and power density on dynamic section shows the influence of shapes and sizes of the geometry of effective MEA regions.

The results obtained in the study show that the fuel cell performance is a function of membrane-active area size and shapes; current collectors opening area size and shapes. The experimental results indicate that with equal areas, the cell performance is best using square shape compared to circle, rhombus, and rectangle. However, in the equal perimeters, the cell performance is best using square shape compared to circle, rhombus, and rectangle.

BACKGROUND

Passive Direct Methanol Fuel Cell (PDMFC) is appropriate for the most range which includes portable and stationary electric & electronics applications as it is a better promising and emerging one, among the power sources in renewable energies. The performance of PDMFC hydrogen ion or proton-conducting electrolyte is dependent totally on polymer electrolyte membrane fuel cell technology.

Direct Methanol Fuel Cell which is portrayed as a gadget that changes over substance vitality into electrical vitality is a promising force source that relies upon the layer and the electro-impetus. The serious issues of the PDMFC are the methanol traverse which discourages cathode performance and declines the fuel effectiveness; this issue can be focused through the advancement of new impetus.

PRIOR ART SEARCH

In PDMFC, a proton-leading polymer membrane covered with impetus on the two sides is applied. This unit is known as a membrane electrode assembly. Polymer electrolyte membrane fuel cells use a strong particle conductive polymer as an electrolyte. Not only the proton conductivity but has it also made physical division of the anode and cathode chambers.

Electric protection and low electrical impedance are the principal significant properties of the electrolyte. The most widely recognized membranes have polymeric perfluoro sulfonic functional derivates, which are fabricated with essential additives for conducting ions.

To analyze the best performance and to find out the highest power density of cell using various shapes by altering gasket openings of dynamic regions with equal areas and equal perimeters. Comparison of the shapes such as square, rectangular, rhombus, and circular active area forms and by using Nickel as current collector material is selected for evaluation.

LITERATURE REVIEW

Vasco S. Silva1, et al. analyzed on characterization, experimentation, and modeling basis. A significant investigation is carried out on essential characteristics of the PDMFC system that focuses primarily on the membrane. This paper focuses chiefly on the PEM performance with an outline of research advancement. Detailed inputs are committed to researching related aspects over the preparation of membrane, its characterization, testing including modeling.

R. Govindarasu in their publication on "Recent evolutions in modeling of direct methanol fuel cell", stated that Direct Methanol Fuel Cells produce one of the better potential substitutes to the substitution of non-renewable energy sources. Anyway, whatever the efficient plan of the cell might be, the cell can't achieve success without legitimate modeling and simulation. The significance and the requirement for the modeling of a DMFC are talked about in detail. The modeling of the key segments, for example, the dispersion layer, MEA, stream dissemination, impetus thickness which add to the performance of the fuel cell are talked about.

Ritesh Pathak presents an analysis of direct methanol fuel cell (DMFC-work and DMFC-froth) frameworks interfaced with DC-DC converters. The yield voltage attributes of six frameworks are thought about. Exploratory outcomes show that DMFC frameworks present variable, nonstraight impedance as a capacity recurrence. It is appeared, through test results, that DMFC frameworks don't display unsurprising yield as the capacity of obligation cycle alone. Therefore, yield voltage control through heartbeat width regulation (PWM) procedures can't be utilized for DMFC frameworks. The inspiration for this analysis is a primer examination concerning the utilization of DMFCs as a force hotspot for a model Martian Uninhabited Aerial Vehicle (UAV).

Jing-Yi Chang, Yean-Der Kuan, and Shi-Min Lee carried out an Experimental Investigation of a Direct Methanol Fuel Cell with Hilbert Fractal Current Collectors. Fractal Current Collectors are fabricated with current collectors of various free opening perimeters and opening ratios. They had experimented on the investigation into the performance of PDMFC as a function of the opening perimeter and on the free open ratio of conducting bipolar plates. But these current collectors could not show better performance.

R. Hashemi & S. Yousefi& M. Faraji in their experimental study of the effect of the active area on the performance of the cell, have observed the methanol consumption over various active areas. Over the huge dynamic zones, as current density is lesser when compared to the lesser dynamic zones, fuel is absorbed gradually. Hence, the local fuel concentration over the polymer membrane is more, resulting in further fuel crossover.

OBJECTIVES

To analyze the best performance and to find out the highest power density of cell using various shapes by altering gasket openings of dynamic regions with equal areas and equal perimeters. Comparison of the shapes such as square, rectangular, rhombus, and circular active area forms and by using Nickel as current collector material is selected for evaluation.

- The objective of the invention is to an Experimental Investigation on Performance of Passive Direct Methanol Fuel Cell using dynamic active regions of Membrane Electrode Assembly with Nickel Current Collectors concerning effective equal areas and equal perimeters by masking membrane with varying shapes and sizes of gasket openings such as square, circle, rhombus, and rectangle.
- The other objective of the invention is to know the performance of the cell with the opening shape of MEA with equal areas and equal perimeters in terms of power densities and current densities.
- The other objective of the invention is the Comparison of gasket opening perimeter and area of effective Opening Shapes with equal areas and equal perimeters.
- The other objective of the invention is the Influence of gasket opening area to perimeter ratio (A/P) on effective gasket opening Shapes with equal areas and equal perimeters.
- The other objective of the invention is the Influence of the Current collector's opening perimeter and opening area on gasket effective Opening Shapes with equal areas and equal perimeters.
- The other objective of the invention is the Influence of the Current collector's opening area to opening perimeter ratio with equal areas and equal perimeters.
- The other objective of the invention is the Influence of the Current collector's opening ratio with equal areas and equal perimeters.
- The other objective of the invention is the Comparison of Power and Power Density on various Opening Shapes and sizes.

DIAGRAM

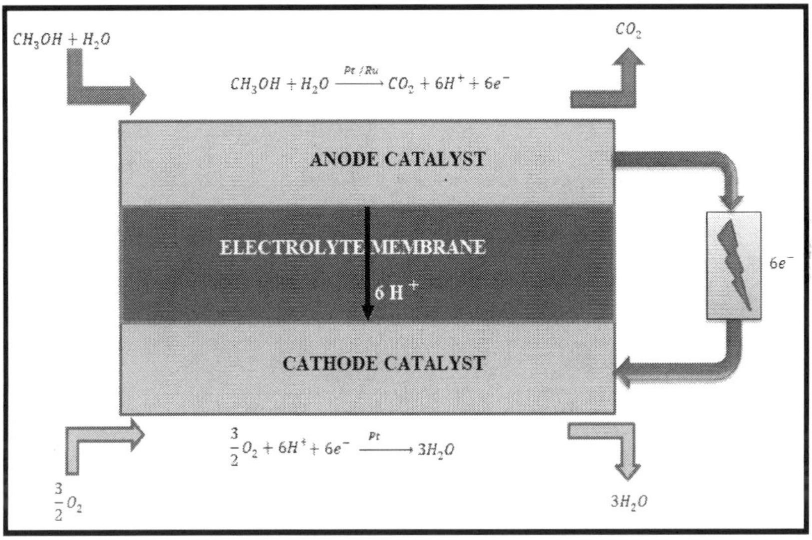

Fig. 1: Schematic presentation of the direct methanol fuel cell

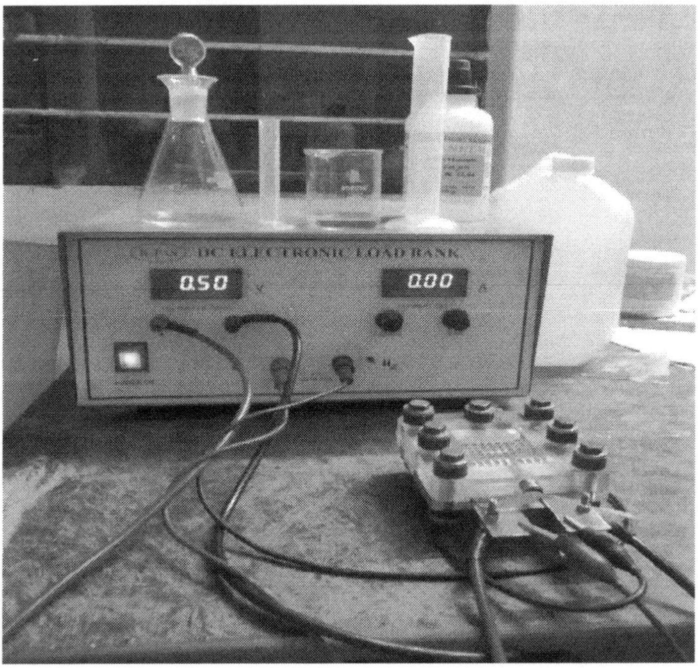

Fig. 2: Experimental Setup of PDMFC

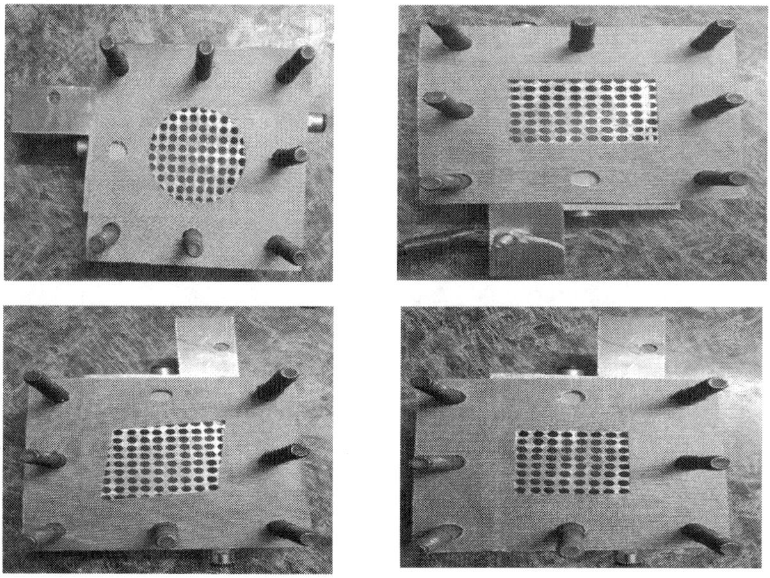

Fig. 3: Setup of the gasket over the anode-side current collector

Fig. 4: Setup of the gasket over MEA

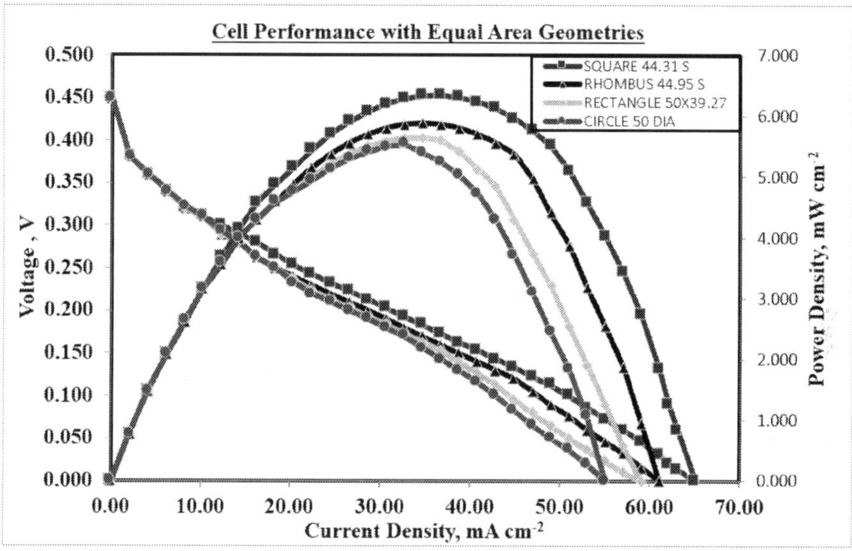

Fig. 5: Cell characteristics with various shapes of equal areas in gasket openings

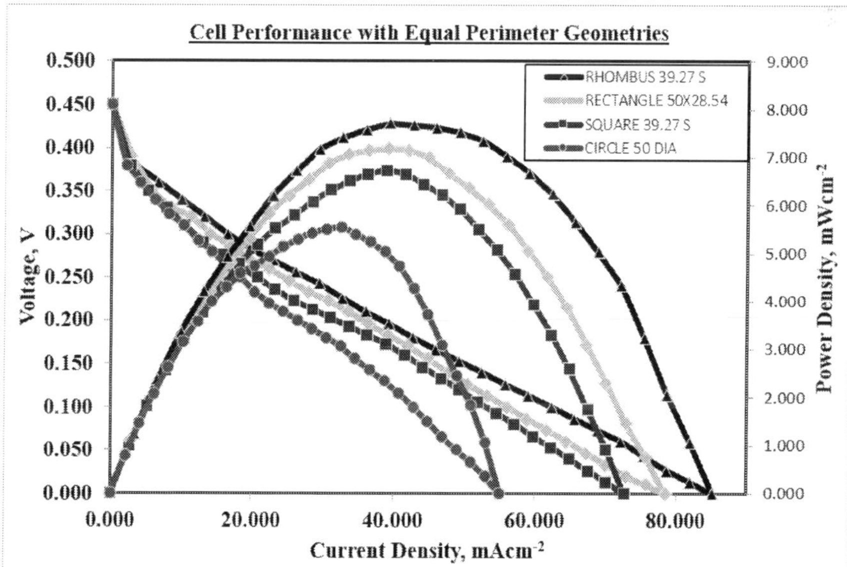

Fig. 6: Cell characteristics with various shapes of equal perimeters in gasket openings

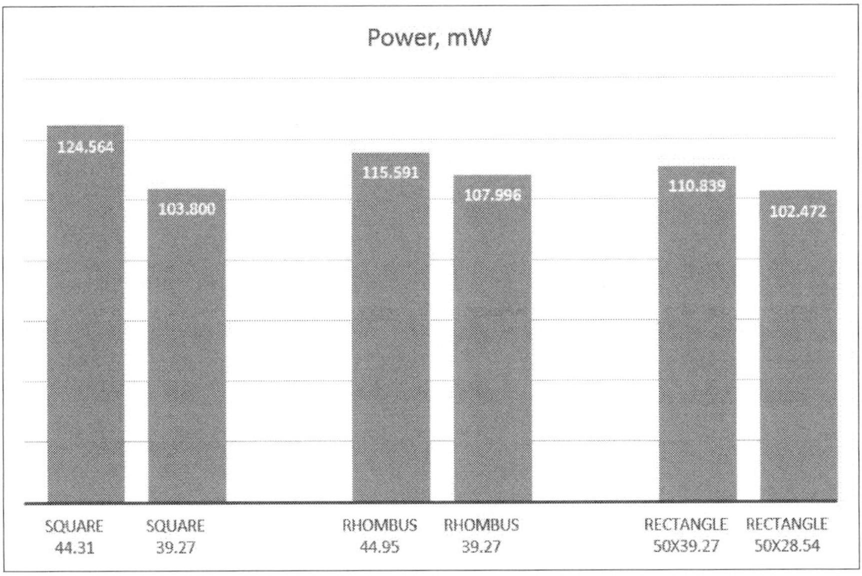

Fig. 7: Power produced by various shapes and sizes in gasket opening.

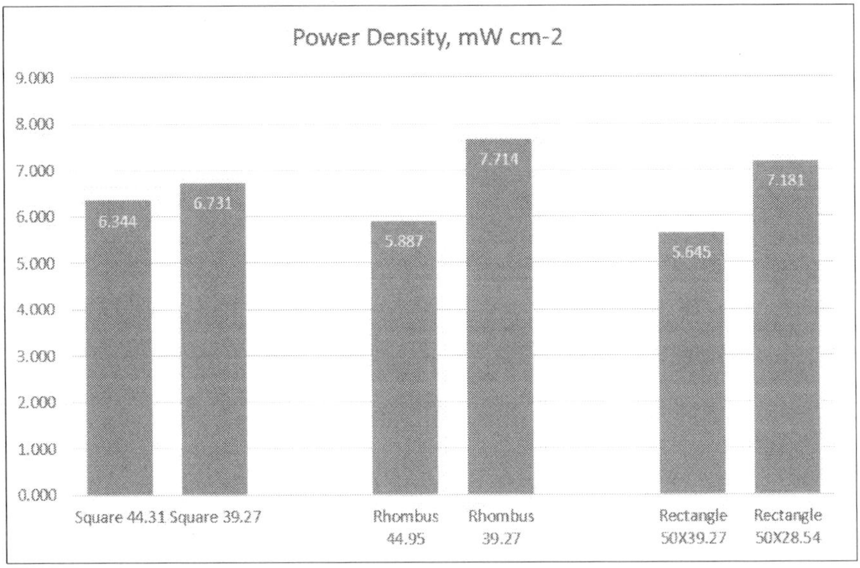

Fig. 8: Power density (bottom) produced by various shapes and sizes in gasket opening.

DESCRIPTION OF THE INVENTION

The essential PDMFC is included by electrical terminals namely anode & cathode, with a strong electrolytic membrane in the middle. The typically used impetuses (catalyst) in PDMFC anode & cathode are Pt/Ru (~4 mg/cm2) and Pt (~2 mg/cm2) respectively. Relating to the electrolyte, PDMFC utilizes a proton exchange partition that isolates the anode and cathode; also permits the movement of hydrogen ions. Even though the essential thermodynamic qualities such as reversible oxidation potential and the methyl alcohol electro-oxidation potential, the mobilization of electrons to the anodic end for fuel conversion to carbon dioxide is also important. The comprised cell processes are shown as anodic end reaction equation (1), cathode end reaction as equation (2) and with overall reaction as equation (3). The schematic cell is shown in figure-1.

$$CH_3OH + H_2O \xrightarrow{Pt/Ru} CO_2 + 6H^+ + 6e^- \quad (1)$$

$$1.5\ O_2 + 6H^+ + 6e^- \xrightarrow{Pt} 3H_2O \quad (2)$$

Overall Fuel Cell Reaction

$$CH_3OH + 1.5\ O_2 \rightarrow CO_2 + 2H_2O \quad (3)$$

EXPERIMENTAL SETUP

To investigate the performance of Passive Direct Methanol Fuel Cell using gaskets of different geometries having Equal Areas and Equal Perimeters with experimental evaluation, a single direct methanol fuel cell fixture is used with current collectors that fabricated using Ni material made with 2 mm thick plate. The current collector material of PDMFC is one of the vital elements that affect the cell performance characteristics. Besides its material for fabrication, design aspects, dimensions with shape factors also influence the cell performance. The essential characteristics of the current collector materials in PDMFC are

- Better electrical conductivity.
- Optimum thermal conductivity.
- High tensile and flexural rigidity of the material.
- Ease of fabrication and machinability.
- Corrosion resistance.

- Longer durability.
- Less density.
- Ease of availability
- Low contact resistance and impedance.
- Even transport and distributing over an active region of reactants.

This experimental setup of the PDMFC is shown in Figure-2. Circular openings of 100 numbers, in 10 by 10 matrix patterns are made with a 3.8 mm diameter drill in the current collector. Openings percentage on current collectors is 45.3% in an effective 25 cm2 area. A solid permeable membrane, Nafion-117 polymer electrolyte is used in MEA. The anode side is dopped with a catalyst around 4 mg cm-2 made up of Platinum-Ruthenium (in equal proportions) on Carbon cloth and on the cathode side, a catalyst of 2 mg cm-2, i.e., Platinum on Carbon cloth is loaded. To avoid leakages of methyl alcohol solutions, PTFE fabricated gaskets are used in between sub-assemblies joining of the DMFC system. An MEA of 5 cm by 5 cm active area is used in the cell and the cell has experimented with 5 molar methanol concentration. The necessary compact clamping is achieved using bolting of M8 size and the uniform torquing is ensured using a handheld torque wrench to a pre-set value of 5 N-m.

Experiment Methodology

Nickel current collector material has been selected for experimentation as it has good electrical conductivity and is better compatible with methanol solution. PTFE gaskets of 0.11 mm thickness are chosen for fabrication of controlling shapes and sizes of effective flow path area over MEA in the cell. The controlling shape of the circular opening in both equal areas and equal perimeters is chosen as a reference. The maximum size circular shape with a diameter of 50mm is cut in the PTFE gasket to accommodate in 50mm side square area of MEA. This circular shape has an effective area of 1963.5 mm2 and a perimeter of 177.2 mm. Keeping this effective area and perimeter as controlling parameters, other shapes such as square, rectangle, and rhombus gaskets are used on either side of MEA at a time. Methyl alcohol solution with 5 molar concentration is used to perform this experiment. By changing current parameters (keeping fixed current values), voltage values are recorded from OCV to the lowest attainable reading. Performance characteristics of the cell are investigated using Polarization and power density curves at this5molar methyl alcohol concentration. The total experiment has been performed thrice to have consistency and repeatability in the noted readings. The arithmetic average reading of the noted values of the experiment conforming to generated voltage and corresponding current is referred for analyzing the cell performance characteristics.

Details of the gasket dynamic openings, such as circle having diameter 50 mm, rectangle length 50 mm X breadth 39.27 mm, square with 44.31 mm side and rhombus with 44.95 mm side in equal areas and the gasket openings such as circle having diameter 50 mm, rectangle length 50 mm X breadth 28.54 mm, square with 39.27 mm side and rhombus with 39.27 mm side in equal perimeters.

The setup of the gasket dynamic openings over the anode-side current collector and over MEA is shown in figure-3 and figure-4 respectively.

SUMMARY OF THE INVENTION

Influence of Effective Opening Shapes on Cell Performance

With equal areas

Polarization (Voltage versus Current density) curves and power density (Power density versus Current density) curves are drawn with the experiment values. It is observed that the cell using square shape has produced a maximum power density of 6.344 mWcm-2 among the other shapes chosen that includes rhombus, rectangle, and circle. The circular shape is producing the least power density of 5.541 mWcm-2. The better performance of the cell with square shape active area is because of equal circulation of reactants on the effective area. Figure-5 has shown the cell characteristics with various shapes of equal areas in gasket openings.

With equal perimeters

Polarization (Voltage versus Current density) curves and power density (Power density versus Current density) curves are drawn from the experimental results. It is observed that the cell with a rhombus shape has produced a maximum power density of 7.714 mWcm-2 among the other shapes chosen that includes square, rectangle, and circle. The circular shape is producing the least power density of 5.541 mWcm-2. The better performance of the cell with rhombus shape is due to less effective area among equal perimeters of selected shapes. Figure-6 has showed the cell characteristics with various shapes of equal perimeters in gasket openings.

Influence of Effective Opening Shapes on Current Density

With equal areas

From the polarization (Voltage versus Current density) curves that are drawn using the experiment values, it is observed that the cell using square shape has produced a maximum current density of 65.2 mAcm-2 among the other shapes chosen that include rhombus, rectangle, and circle. The circular shape is producing the least current density of 55.0 mAcm-2. The better performance of the cell with square shape active area is because of equal circulation of reactants on the effective area.

With equal perimeters

Polarization (Voltage versus Current density) curves are drawn from the experimental results, it is observed that the cell with rhombus shape has produced a maximum power density of 85.3 mAcm-2 among the other shapes chosen that include square, rectangle, and circle. The circular shape is producing the least power density of 55.0 mAcm-2. The better performance of the cell with rhombus shape is due to less effective area among equal perimeters of selected shapes.

Comparison of Gasket Opening Perimeter and Area Effective Opening Shapes

With equal areas

From the various shapes of equal areas, it is observed that the circular shape has the lowest opening perimeter and the rhombus has the largest perimeter. Followed by rhombus shape, rectangle and square shapes have fewer opening perimeters in the descending order.

With equal perimeters

From the various shapes of equal perimeters, it is observed that the rhombus shape has the lowest opening area and the circle has the largest area. Therefore, it is measured from the geometries of various shapes that rhombus, rectangle, square and circle have increasing gasket opening areas. Hence power density is influenced by the area of the opening in equal perimeter shapes.

Influence of Gasket Opening Area to Perimeter Ratio (A/P) on Effective Gasket Opening Shapes

With equal areas

From the various shapes of equal areas, it is observed that the ratio of gasket opening area to the perimeter is also influencing the performance. The circular shape has the highest opening area to the perimeter and the rhombus has the least value. From the performance graphs, it is revealed that the circular one has the least performance due to the influence of the highest gasket opening area to perimeter ratio.

With equal perimeters

From the various shapes of equal perimeters, it is observed that the ratio of gasket opening area to the perimeter is also influencing the performance. The circular shape has the highest opening area to the perimeter and the rhombus has the least value. From the performance graphs, it is revealed that the circular one has the least performance due to the influence of the highest gasket opening area to perimeter ratio and the rhombus shape has the highest performance due to the lowest gasket opening area to perimeter ratio. It is revealed that the gasket opening areas with various shapes such as rhombus, rectangle, square, and circle have better performance with decreasing gasket opening area to perimeter ratio on effective opening shapes.

Influence of Current Collector's Opening Perimeter and Opening Area on Gasket Effective Opening Shapes

With equal areas

From the selected shapes in equal areas, it is observed that the current collector opening perimeter is influencing the performance of the cell. The rectangular shape has the highest opening perimeter and the square has the least value. From the performance graphs, it is revealed that the square shape has the highest performance due to the influence of the least opening perimeter on the current collector.

Similarly, from the various shapes of equal areas, it is observed that the current collector opening area is also influencing the performance. The square shape has the largest opening area and the rectangle has the least value. From the performance graphs, it is identified that the square shape has the highest performance due to the influence of the least opening area on the current collector.

With equal perimeters

From the identified shapes of equal perimeters, it is observed that the current collector opening perimeter is also influencing the cell performance. Rhombus shape has the least opening perimeter and circular shape has largest opening perimeter on the current collector. From the performance graphs, it is observed that the circular one has the least performance due to the largest opening perimeter on current collectors and the rhombus shape has the highest performance due to the lowest opening perimeter on current collectors.

Similarly, from the various shapes of equal perimeters, it is observed that the current collector opening area is also influencing the performance. Rhombus shape has the least opening area and the circle has the largest opening area on current collectors. From the performance graphs, it is revealed that the rhombus shape has the highest performance due to the influence of the least opening area on the current collector.

Influence of Current Collector's Opening Area to Opening Perimeter Ratio

With equal areas

From the various shapes of equal areas, it is observed that the square shape has the largest ratio of current collector opening area to the perimeter, whereas the rectangle has the least ratio. Besides other influencing factors, the square shape is showing better performance with the largest ratio of current collector opening area to the perimeter.

With equal perimeters

From the different shapes of equal perimeters, it is observed that the rhombus shape has the least current collector opening area to perimeter ratio and the rectangle has the largest ratio. Besides other consideration factors, rhombus shape is showing the better performance with the least ratio of current collector opening area to the perimeter.

Influence of Current Collector's Opening Ratio

With equal areas

From the assorted shapes in equal areas, it is observed that the square shape has the largest current collector opening ratio and the rectangle has the least ratio. Besides other factors, square shape is showing better performance with the largest ratio of current collector opening ratio.

With equal perimeters

From the various shapes of equal perimeters, it is observed that the rhombus shape has the least current collector opening ratio and the rectangle has the largest ratio. Besides other factors, rhombus shape is showing the better performance with the least ratio of current collector opening.

Comparison of Power and Power Density on Opening Shape Sizes

Power generated Vs various sizes and shapes

From the assorted sizes of selected shapes, a comparison study on size factor (in terms of area of the dynamic zone) with Power produced is carried out. It is observed that the higher the effective MEA area, the higher the power produced by the cell for a given shape.

Power Density produced Vs various sizes and shapes

From the various sizes of selected shapes, a comparison study on size factor (in terms of area of the dynamic zone) with Power density produced is carried out. It is observed that the higher the effective MEA area, the lower the power produced by the cell for a given shape. Hence the performance of the cell lowers with enhance in the size of the dynamic zone.

Figure-7 and figure-8 have shown the comparison of power and power density respectively produced against the gasket opening area with various shapes and sizes.

OUTCOME

Our Invention is related to an Experimental Investigation on Performance of Passive Direct Methanol Fuel Cell using dynamic active regions of Membrane Electrode Assembly with Nickel Current Collectors. Comparison of the shapes such as square, rectangular, rhombus, and circular active area forms and by using Nickel as current collector material is selected for evaluation. In the equal areas with gasket opening area of 1963.5 (+/-0.2) mm2, among circular, rhombus, rectangular and square shapes chosen for investigation on the performance of cell at 5 molar methanol concentration, square shape opening consisting of the perimeter of 177.2 mm has developed highest power density of 6.344 mWcm-2with a corresponding maximum current density of 65.2 mAcm-2. In the equal perimeters of 157.1 (+/- 0.2) mm of gasket openings, among circular, rhombus, rectangular and square shapes chosen for investigation on the performance of cell at 5 molar methanol concentration, Rhombus shape opening consisting of area 1400 mm2has developed the highest power density of 7.714 mWcm-2 with a corresponding maximum current density of 85.3 mAcm-2.

- According to claim 1# the invention is related to an Experimental Investigation on Performance of Passive Direct Methanol Fuel Cell using dynamic active regions of Membrane Electrode

Assembly with Nickel Current Collectors In the large active areas, methanol consumption is slow because the current density is lower when compared to the smaller active areas. As a result, the surface-specific concentration on the ion-conducting membrane is being more, making a higher crossover of methanol over bigger active zones.

- According to claim 1,2# the invention is the experimental investigation demonstrated that the size of the dynamic zone significantly affects the performance of the cell and its performance can be improved with the expansion in dynamic region size. In the vast dynamic regions, the current density is lower compared to the smaller dynamic zones where the methanol is expended gradually. So, the local concentration on the cell membrane is more and brings further fuel cross-over.
- According to claim 1,2,3# the invention is the exothermic response that is involved on penetrated fuel and oxidant on cathode impetus generates more heat, which prompts an elevated working temperature. This elevated working temperature brings about superior electrochemical kinetics of fuel oxidation and a decrease in oxygen response, leading to superior cell performance.
- According to claim 1,2,3,4# the invention is the internal obstruction decreases with expanding the size of the dynamic zone. This reality is good with the accomplished outcome. Even though the cell performance can be enhanced by expanding the size of the dynamic zone, with the research outcome as on date, it is revealed that tiny dynamic zones have its focal points, for example, more slow charge transfer obstruction, lesser flooding, even distribution of methanol dispersion, increasing of effective current gathering, minor expenditure on fabrication, and further, it is small in size and weight.
- According to claim 1,2,3,5# the invention is from the experiment that cell performance depends on the combined effect of (i) gasket opening shape, opening area, and its perimeter; (ii) the free open ratio and opening perimeter on the current collector plates. Comparison study referring to power and power density on dynamic section shows the influence of shapes and sizes of the geometry of effective MEA regions.

- According to claim 1,2,3,6# the invention is that the fuel cell performance is a function of membrane-active area size and shapes; current collectors opening area size and shapes. The experimental results indicate that with equal areas, the cell performance is best using square shape compared to circle, rhombus, and rectangle. However, in the equal perimeters, the cell performance is best using square shape compared to circle, rhombus, and rectangle.

CITATIONS

- Munawira Kotyad (Associate Professor), Yvette Lee, (Social Worker and Counsellor), 202121052144, MAHATMA EDUCATION SOCIETY'S, PILLAI HOC COLLEGE OF ENGINEERING AND TECHNOLOGY, "Indian Child Specific Game Based Psychotherapeutic (therapy for post-traumatic stress disorder" 2021, Indian Patent office.
- Sonam Prasad Bhopatrao, Prof (Dr.) Shilpa Pankaj Kewate, 202121052897, MAHATMA EDUCATION SOCIETY'S, PILLAI HOC COLLEGE OF ENGINEERING AND TECHNOLOGY, " Sustainable Sandwich Composite Circular Manhole Cover Designed for Pedestrian Networks" 2021, Indian Patent office.
- Dr. Shabab Ahmed Rizvi (Assistant Professor), Dr. Gajanan Wader (Principal), 202121052893, MAHATMA EDUCATION SOCIETY'S, Pillai College of Arts Commerce & Science, Dr K M Vasudevan PILLAI Campus, Sector 16, New Panvel," E-O Indicator: Digital Engine oil indicator shows accurate conditions of engine oil" 2021, Indian Patent office.
- Nisha Subhash Yadav, 202121052895, MAHATMA EDUCATION SOCIETY'S, Pillai College of Arts Commerce & Science, Dr K M Vasudevan PILLAI Campus, Sector 16, New Panvel, NAVI Mumbai -410206, MH, India." Wireless Electric Current Transmitter and Receiver for Electronic Devices." 2021, Indian Patent office.
- Shardul Buva, Dr. Gajanan Pandit Wader, 202121053764, , MAHATMA EDUCATION SOCIETY'S Pillai College of Arts Commerce & Science, Dr K M Vasudevan PILLAI Campus, Sector 16, New Panvel, NAVI Mumbai -410206, MH, India, " Vehicle Spare Parts Identification System ",2021, Indian Patent office.
- Rohit Devappa Gawade, Prof(Dr.) Shilpa Pankaj Kewate,202121054010, MAHATMA EDUCATION SOCIETY'S, PILLAI HOC COLLEGE OF ENGINEERING AND TECHNOLOGY, "Manufacturing of lightweight sandwich bricks using a unique combination of Bamboo Fiber and E-waste Powder" 2021, Indian Patent office.
- Dr. Sundeep Harish Deulkar, Dr. Arun Pillai, Dr. Malavika P. Sharma, 202121053765, MAHATMA EDUCATION SOCIETY'S, PILLAI COLLEGE OF ENGINEERING, "Use of Kinaesthetic Game Based Learning for Modelling the Hypothetical Energy Levels of A LASER", 2021, Indian Patent office.

Chapter - 06

IoT Based Water Quality Monitoring System Using ML

Jayesh Rane

MES's, Pillai HOC College of Engineering and Technology, Rasayani, Maharashtra India.

Mansi Subhedar

MES's, Pillai HOC College of Engineering and Technology, Rasayani, Maharashtra India.

Pooja Kulkarni

MES's, Pillai HOC College of Engineering and Technology, Rasayani, Maharashtra India.

ABSTRACT

Water pollution has been a real threat to living species. The real-time monitoring of drinking water is nothing less than a challenging task. The purpose of this project is to develop a low-cost system for the real-time monitoring of water quality using Machine Learning (ML). Some of the important parameters of water such as pH, turbidity, conductivity and water temperature are measured using respective sensors and Water Quality Index (WQI) is calculated. Arduino controller is employed to process the measured value from the sensors. The data acquired from Sensors are sent to the Flask server which is used to create a webpage interfaced with the Machine Learning model to upload sensor values and predicted water quality. Support Vector Machine (SVM) and Random Forest (RF) algorithms are used in the analysis and prediction of water quality.

INTRODUCTION

Nowadays, lots of new technologies have emerged but at the same time, we are facing problems of global warming, pollution, shortage of natural resources, etc. The fast growing technology is affecting different natural resources like water very badly. Diseases such as typhoid, diarrhoea, and cholera are caused due to contaminated water. According to reports from WHO, around 21% of diseases are caused because of the usage of such polluted water and 77 million people are affected by these diseases [1]. Water is the most essential element for the entire humanity. Hence its safety and quality should be a major concern.

The traditional way of collecting water samples and examining them manually in the laboratories is a time taking process. These types of techniques are not suitable for real-time observation. With IoT and Machine Learning algorithms, it is easy to obtain the sensor values from a water sample and predict the quality of water in the comfort of our home. The sensors are placed in the water to be tested. This water can be either stored water or running water. Sensors convert the physical parameter into an equivalent measurable quantity. This can be given as input to the controller. IoT facilitates sensors to transfer data between them or to the cloud without the intervention of humans. The water quality index of the water, which helps in determining the quality of water, can be predicted by the extensive use of a machine learning regression algorithm. Here two Machine learning algorithms viz Random Forest (RF) and Support Vector Machine (SVM) are employed for classification purposes.

Spandana et al. employed Node MCU -WIFI module and sensors to collect the real-time data. When the sensors are in peculiar conditions, the alarm rings [10]. They projected the new approach of IoT based water quality monitoring in real-time. V Vaishnavi et al proposed a low-cost system which can automatically monitor water quality. Several sensors were used to measure the physical and chemical parameters like temperature, PH, turbidity etc. of the water. The measured values from the sensors were given as inputs to the core controller for further processing. The sensor data can be viewed on the internet using a WI-FI system [9].

Gokulanathan, et. al. presented a new approach of water quality monitoring through a wireless sensor network. In case of any deviation of water quality parameters, the system provides an alarm to the remote user. Three important parameters such as pH, temperature and dissolved oxygen are monitored in this system. The diseases caused due to the presence of metals and pollutants in the water can be protected by this system [8]. Lambrou et al. used two transceivers and a transmitter at the place where they wanted to observe the water quality. They have proposed a reliable, cost-effective and portable water level control system. The two transceivers used work on RF frequencies and are used for wireless communication to the internet server [7]. Pasika et. al. developed an efficient, inexpensive IoT solution for real-time water quality monitoring. They have proposed an efficient algorithm to keep track of the quality of water in real-time. They have used ThingSpeak to observe temperature and humidity, the pH value, turbidity of the water, and the level of water in the tank through the webserver. These parameters can also be seen on ThingSpeak mobile application [11].

Umair Ahmed et al came up with an alternative approach of using supervised machine learning to predict the quality of water and achieved a reasonable accuracy using a minimal number of parameters [19]. Bharath Singh J et al designed a system to predict the quality of water using Machine learning based on the water standards suggested by BIS to prevent the diseases that are caused due to poor water quality. They have built a Machine Learning model and have integrated it with the webpage where sensor values are displayed, to get an instant alert of the quality of water [3].

The proposed system consists of various sensors to monitor and upload the values to the webpage. Machine learning algorithms are used to predict the quality of water by analysing the parameter values given by the sensors. The ML algorithms have integrated with the webpage where the sensor values are displayed. So that the user can come to know the quality of water. Flask framework is used to develop a webpage.

System Requirements

The various sensors used in this work are explained in the following section.

i) pH sensor

The pH of hydrogen ion concentration is negative in nature. An acidity balancing test or the alkaline content of hydrogen ions in water is used for the calculation [12]. The pH of water sources is around 7, and the pH varies from 6.5 to 9.5, which is considered acceptable drinking water [15]. Acidic solutions have a pH of 0 and alkaline solutions have a pH of 14. The concentration of hydrogen ions falls with an increase in multiple pH levels, and water becomes less acidic. A pH sensor contains a measuring electrode and a reference electrode. The hydrogen ion is sensitive to electrode measurements with a voltage that is proportional to the ion charge density in the hydrogen solution. Because the electrical differential potential is temperature-dependent, a temperature sensor is required to adjust the voltage change. [13].

ii) Turbidity sensor

Turbidity is a measurement of the cleanliness of water or the amount dispersed in it. It evaluates light transmittance and dispersion rate by using light to identify suspended particles [11]. The estimate takes into account the number of water particles floating in the water, such as plant waste, dust, silt, and mud, which affect the amount of sunlight that reaches the water [14]. Increased turbidity can harm marine life and cause a variety of illnesses in humans. The rate varies depending on how many particles are suspended in the water. With rising turbidity, the amount of total suspended solids (TSS) in the water will increase. Both digital and analogue outputs are produced by the sensor. The sensor has a 5V input voltage and a 0 to 4.5V analogue output voltage. It can endure temperatures between 100 and 900 degrees Fahrenheit. Its measurement unit is NTUs (Nephelometric Turbidity Units). The sensor is, in essence, placed to the side of the beam. If many tiny particles are scattered in the water when light reaches the sensor, the source beam will identify these small particles.

iii) Temperature Sensor

The DHT11 is a standard device for measuring the temperature and humidity of the external environment. The sensor contains a Negative Temperature Coefficient (NTC) for temperature measurement and an 8-bit microscope for output, resulting in serial data for temperature and humidity measurements transmitted to the microcontroller. The sensor's temperature and humidity values are identical. It is also properly calibrated, so it does not need to be calibrated again and is therefore simple to interface. The sensor can determine temperatures ranging from 0°C to 50°C, as well as humidity levels ranging from 20% to 90%, with an accuracy of 1°C to 1% [13]. The sensor determines the temperature of the atmosphere which is essential for the pH and turbidity sensors to function properly over time. The types of marine organisms that can live in the sea can also be determined by temperature measurements. [14]

iv) Water Conductivity Sensor

The ability of a liquid to transmit an electric current is measured by its conductivity. The ability of a water sample to conduct an electrical current is measured using conductivity sensors. The conductivity of a sample is increased by adding ions from dissolved salts and inorganic elements such as alkalis, chlorides, sulphides, and carbonate compounds, making conductivity a significant water quality assessment.

v) Dissolve Oxygen Sensor

The amount of dissolved oxygen in a unit of water is referred to as dissolved oxygen (DO). Oxygen enters water by diffusion in the surrounding air, aeration (turbulent movement), and as a waste product from photosynthesis in plants.

An anode and a cathode are used in DO sensors. These electrodes are submerged in electrolytes on both sides (inside the sensor body). The anode and cathode are separated from the measured water by an oxygen-permeable layer. The permeable membrane allows oxygen from the sample water to enter the sensor and be lowered at the cathode.

This chemical process generates an electrical signal, which is sent from the cathode to the anode and subsequently to the dissolved oxygen sensor. The pressure changes across the membrane caused by oxygen consumption at the cathode changes depending on the partial pressure of oxygen in the sample. As a result, the oxygen concentration rises and the partial pressure and rate of diffusion rise as well.

Simulation Requirements

i) Flask Server

One of the most widely used Python web application frameworks is Flask. It's a WSGI web application framework that's lightweight on resources. The framework was built with the capability of scaling up to more sophisticated applications in consideration.

ii) Random Forest Algorithm

When given a set of predictor values, the Random Forest (RF) algorithm creates an ensemble of simple tree predictors, each capable of delivering a response. For classification problems, the result might be a class membership, which connects or classes a set of independent predictor values to one of the variable's categories. The outcome is determined by an arbitrary number of simple trees. In the case of classification problems, an ensemble of simple trees selects the most popular class. When tree ensembles are used, prediction accuracy is significantly improved. Every tree's output is determined by a set of predictor values chosen independently (with replacement) and the forest's tree distribution, which is a subset of the original data set's predictor values. Log2M+1 is the ideal subset size for predictor variables, where 'M' is the number of inputs. The RF's predictions are considered as the average of the trees' predictions. Random Forest Prediction:

kp =1/k Kth tree response

Where the 'k' runs across the forest's individual trees. In the predictor variables, RFs may include missing data in a variety of ways. If a predictor variable is selected at the root node at a time in the tree sequence for which certain instances have no valid data, the prediction for those cases is simply based on the overall mean at the root node. As a result, neither the elimination of cases with missing data predictors nor the computation of split statistics are required. In Python, you may import the Random Forest Regressor from the SCI-KIT learn package. We've decided on a value of 10 for our n-estimators (number of decision trees). The model was given the training data and predicts the Water Quality Class.

iii) Support Vector Machine Algorithm

Support vector machine is a prominent Supervised machine learning technique for solving problems of classification, regression, recognition, and time series analysis (Kara et al., 2011). However, it is most commonly used in classification problems.

Because of qualities like a strong generalisation, high prediction accuracy, and quicker training time, SVM has become relatively popular. SVM has a great capacity to categorise both linear and non-linear data and can operate flawlessly with small sample sets. In the two-classification problem, if there exist some original training samples which are represented as xi= {x1, x2, x3...xn} in the input space where xi = Rd and their corresponding labels are represented as yi = {-1, +1} (Kara et al., 2011). The basic purpose of SVM is to create a hyperplane between training samples that divides them by a maximum margin, allowing them to be mapped from input space to higher dimensional feature space. As a result, the +1 labelled vector is on one side of the hyperplane, while the -1 labelled vector is on the other.

Design Methodology

Figure 1 shows a block diagram of the proposed system. The parameters used to assess the water quality include DO, temperature, pH, turbidity, and conductivity.

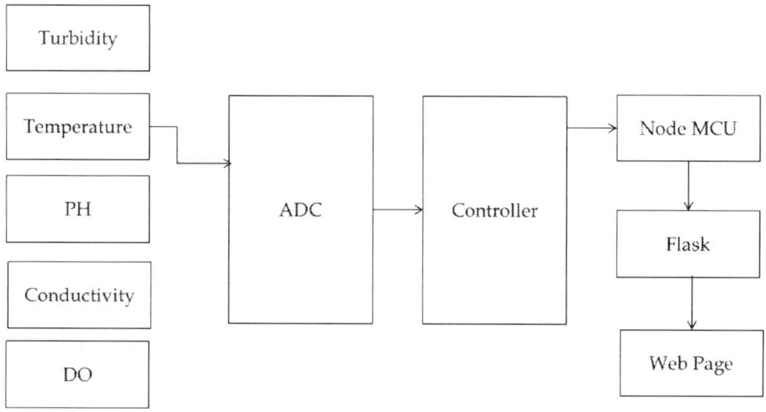

Fig. 1: Block Diagram of Proposed System

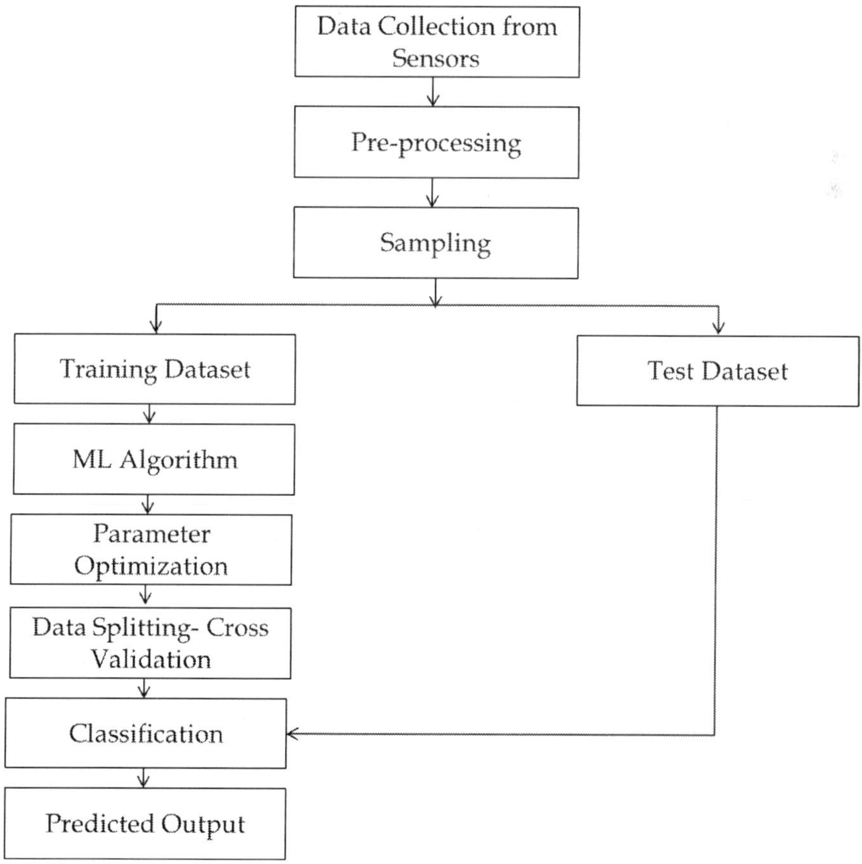

Fig. 2: Proposed Prediction model using Machine Learning

Accordingly, the sensors are used to obtain the values for various water samples. It consists of a Dissolved Oxygen sensor, pH sensor, temperature sensor, turbidity sensor, and Conductivity sensor. The sensor values obtained are analog and are converted to digital using ADC. Arduino is the controller used in work. These values are processed and further employed as an input to the prediction model. The data is sent to the server using Node-MCU, which then shows the results. Because Node-MCU is a WIFI module, it connects to the internet immediately.

The machine learning model used for the prediction of quality is depicted in Figure 2. The readings obtained from sensors are pre-processed. To prepare training and testing datasets, 5 fold cross-validation is employed. The results obtained with both random forest and SVM algorithms were compared.

RESULTS AND DISCUSSION

The accuracy metrics used to assess the machine learning algorithms' performance are discussed as follows.

Accuracy Measures

(1) Mean Absolute Error (MAE)

Mean absolute error (MAE) is a measure of accuracy for regression. It sums up absolute values of errors and divides them by the total number of values. It gives equal weight to each error value. The formula for calculating MAE is shown in Equation (1), where xobs refers to the actual value, xpred refers to the predicted value, and n refers to the total number of samples considered [18].

$$\text{MAE} = \frac{\sum \ (||x_{obs} - x_{pred}||)}{n} \quad ...(1)$$

(2) Mean Square Error (MSE)

Mean square error (MSE) is the sum of squares of errors divided by the total number of predicted values. This attributes greater weight to larger errors. This is particularly useful in problems where there needs to be a larger weight for larger errors. It is measured by Equation (2), where xobs is the actual value, xpred is the predicted value, and n is the total number of samples considered [18].

$$\text{MSE} = \frac{\sum \ (x_{obs} - x_{pred})^2}{n} \qquad ...(2)$$

(3) Root Mean Squared Error (RMSE)

Root mean squared error (RMSE) is just the square root of MSE and scales the values of MSE near to the ranges of observed values. It is estimated from Equation (3), where xobs points to the actual value, xpred points to the predicted value, and n points to the total number of samples considered [18].

$$\text{RMSE} = \sqrt{\frac{\sum \ (x_{obs} - x_{pred})^2}{n}} \qquad ...(3)$$

(4) R Squared Error (RSE)

R squared error (RSE), also known as the coefficient of determination, and often denoted as R2, determines the goodness of fit of the model. It particularly explains the amount of variance of the dependent variable that is explainable through the independent variable, as shown in Equation. Higher RSE values mean that the independent variables largely explain the variance of the dependent variable [17].

$$\text{RSE or R2} = 1 - \frac{Explained\ Variation}{Total\ Variation} \qquad (4)$$

For classification, we used the following measures:

(1) Accuracy

Accuracy is the correct number of predictions made by the model over all the observed values. Accuracy is measured by Equation (5), where TP refers to true positive, TN refers to true negative, FP refers to false positive and FN refers to false-negative [15, 16].

$$\text{Accuracy} = \frac{TP + TN}{TP + FP + TN + FN} \qquad ...(5)$$

(2) Precision

Precision is the proportion of correctly classified instances of a particular positive class out of the total classified instances of that class. Precision is calculated with the formula shown in Equation (6), where TP refers to true positive and FP refers to false positive [15, 16, 17].

$$\text{Precision} = \frac{TP}{TP+FP} \quad ...(6)$$

(3) Recall

A recall is the proportion of instances of a particular positive class that was actually classified correctly. The recall is calculated with the formula shown in Equation (7), where TP refers to true positive and FN refers to false negative [15, 16, 17].

$$\text{Recall} = \frac{TP}{TP+FN} \quad ...(7)$$

(4) F1 Score

As precision and recall, individually, do not cover all aspects of the accuracy, we took their harmonic mean to reflect the F1 score, as shown in Equation (8), which covers both aspects and reflects the overall accuracy measure better. It ranges between 0 and 1. The higher the score, the better the accuracy [15, 16, 17].

$$\text{F1 Score} = \frac{2*Precision*Recall}{Precision+Recall} \quad ...(8)$$

Water Quality Index (WQI) Calculation

A water quality index is a numerical number based on individual metrics that measure the overall quality of the water. The usefulness of a water sample may be measured using the quality index. Prior to computing the water index, standard values for the parameters must be chosen based on WHO/BIS/ICMR recommendations. The Water Quality Index may be calculated in a variety of ways; the Generalized Arithmetic Water Index technique is employed here. pH, total dissolved salts, turbidity, electrical conductivity, and dissolved oxygen were the parameters we have chosen for analysis. The parameters and their standard values are given in the following Table 1:

Table 1: Water Quality Parameters and their values

Parameter Standard	Parameter Standard
pH	6.5-8.5 (BIS)
TDS (mg/l)	<=500 (BIS)
Turbidity (NTU)	<=5 (BIS)
Electrical Conductivity (μS/cm)	<=300 (ICMR)
Dissolved Oxygen	10 mg/L

For estimating WQI, a regression model is employed. The accuracy ratings of several regression models were determined after they were trained. For calculating the WQI, the Random Forest and SVM are used, and their results accuracy comparison is been made. WQI is considered in classification algorithms to identify the water quality class (WQC) of each sample after estimating the WQI. The classes are defined as shown in Table 2.

Table 2: Classes for water quality index

Water Quality Index Range	Class
85 to 100	Very Good
65 to 85	Good
65 to 45	Fair
< 45	Poor

In this work a total of 500 samples are used. To obtain better results, 5-fold cross validation is employed and results are obtained with training and testing sets using both random forest algorithm and support vector machine. The obtained results for these output metrics are presented in Table 3 and Table 4.

Table 3: Regression results using four parameters.

Algorithm	MAE	MSE	RMSE	RSE
Random Forest	1.345	1.236	1.012	1.231
SVM	1.456	1.478	2.345	2.126

Table 4: Classification results using four parameters.

Algorithm	Accuracy	Precision	Recall	F1 Score
Random Forest	97.21	95.23	0.653	98.96
SVM	96.23	94.34	0.545	98.01

From Table 3 and Table 4 results, it can be observed that the random forest-based approach outperforms SVM based algorithm in this work.

CONCLUSION

In this work, IoT based cost-effective water quality monitoring system is proposed using a machine learning technique. The various water quality parameters used to assess the quality includes pH, turbidity, conductivity, DO etc. The sensor readings were collected and processed with Ardunio. The data values were transferred to flask server for further processing and prediction. Random Forest and SVM techniques were employed for the prediction of water quality.

The quality metrics used to assess the model performance shows the random forest based approach gives better predicted output. The system is economical and easy to implement real time mechanism to predict the water quality and can be used in several applications.

REFERENCES

[1] K Gopavanitha and S Nagaraju, A low cost system for real time water quality monitoring and controlling using IOT, International Conference on Energy, Communication, Data Analytics and Soft Computing (ICECDS), DOI.10.1109/ICECDS.2017.8390054 2017

[2] Jcho Zinmyint, Lenin Gopal, and Yan Lin Aung, "Reconfigurable Smart Water Quality Monitoring System in IoT Environment", 978-1-5090-5507 4/17/31.002017IEEEICIS26,2017, Wuhan, China, 2017

[3] Bharath Singh J, Nirmitha S, Kaviya S S, "Smart Urban Water Quality Prediction System Using Machine Learning", International Conference on Recent Trends in Computing (ICRTCE-2021) doi:10.1088/1742-6596/1979/1/012057, 2021

[4] Machine Learning World Health Organization. Guidelines for Drinking-Water Quality, 4th ed.; World Health Organization: Geneva, Switzerland, ISBN 978-92-4-154995-0, 2011

[5] Nikhil Kedia, "Water Quality Monitoring for Rural Areas- A Sensor Cloud Based Economical Project", 1st International Conference on Next Generation Computing Technologies (NGCT-2015) Dehradun, India, 4-5 September 2015. 978-1-4673-68094/15/31.002015IEEE 37

[6] Lambrou, Theofanis P., Anastasiou, Christos C., Panayiotou, Christos G., Polycarpou, Marios M., 2014. A low-cost sensor network for real-time monitoring and contamination detection in drinking water distribution systems. IEEE Sensor. J. 8, 2765–2772.

[7] S Gokulanathan, P Manivasagam, N Prabu and T Venkatesh, "A GSM Based Water Quality Monitoring System using Arduino", Shanlax International Journal of Arts Science and Humanities, vol.6, no. 4, pp. 22-26, 2019

[8] V. D. Vaishnavi and M. A. Gaikwad, "Water Quality Monitoring System Based on IOT, Advances in Wireless and Mobile Communications", vol.10, no. 5, pp. 1107-1116, 2017

[9] Kancheti Spandana and V R Seshagiri Rao, "Internet of Things (IOT) based Smart Water Quality Monitoring System", International Journal of Engineering & Technology, vol. 7, no.3, pp. 259-262, 2018

[10] Sathish Pasika, Sai Teja Gandla, "Smart water quality monitoring system with cost-effective using IoT", Heliyon, vol. 6, Issue 7, 2020, e04096, ISSN 2405-8440, https://doi.org/10.1016/j.heliyon.2020.e04096.

[11] Cloete, Niel Andre, Malekian, Reza, Nair, Lakshmi, "Design of smart sensors for real-time water quality monitoring" Department of Electrical, Electronic and Computer Engineering, University of Pretoria, Pretoria, South Africa. IEEE J. 13, 1–16, 2014

[12] Zin Myint, Cho, Gopal, Lenin, Lin Aun, Yan, "Reconfigurable smart water quality monitoring system in an IoT environment" IEEE ICIS 435–440, 2017

[13] Daigavane, Vaishnavi V., Gaikwad, M.A., "Water quality monitoring system based on IoT" Adv. Wireless Mobile Communication ISSN 10, 1107–1116, 2017

[14] Shafi, U. Mumtaz, R.; Anwar, H. Qamar, A.M. Khurshid, H. , "Surface Water Pollution Detection using Internet of Things", Proceedings of the 2018 15th International Conference on Smart Cities: Improving Quality of Life Using ICT & IoT (HONET-ICT), Islamabad, Pakistan, 8–10 October 2018; pp. 92–96.

[15] Sokolova, M.; Japkowicz, N.; Szpakowicz, S. Beyond accuracy, F-score and ROC: A family of discriminant measures for performance evaluation. In Proceedings of the Australasian Joint Conference on Artificial Intelligence, Hobart, Australia, 4–8 December 2006; pp. 1015–1021.

[16] Goutte, C.; Gaussier, E. A probabilistic interpretation of precision, recall and F-score, with implication for evaluation. In Proceedings of the European Conference on Information Retrieval, Santiago de Compostela, Spain, 21–23 March 2005; pp. 345–359.

[17] Willmott, C.J. Matsuura, K. Advantages of the mean absolute error (MAE) over the root mean square error (RMSE) in assessing average model performance. Clim. Res. 2005, 30, 79–82

[18] Umair Ahmed, Rafia Mumtaz, Hirra Anwar Asad A. Shah, Rabia Irfan and José García-Nieto, "Efficient Water Quality Prediction Using Supervised Machine Learning" Water 2019, 11, 2210; doi:10.3390/w11112210

CITATIONS

- GAURAV GYANENDRA MISHRA, PRAKASHDEV INDRADEO SINGH, DR. SHILPA PANKAJ KEWATE PANKAJ KISHOR MAHALE, SALONEE RANGNEKAR, SHWETANG VASANT MESHRAM, AKASH YADAV, SNEHA SATISH MUSALE, VEDANT SANDEEP SAWANT, DEVESH THAKER, 202121053767, MAHATMA EDUCATION SOCIETY'S, PILLAI HOC COLLEGE OF ENGINEERING AND TECHNOLOGY, " Steel fibre reinforced sustainable composite hollow brick", 2021, Indian Patent office.
- Dr. C K PRASHANT, REMYA MADAN GOPAL, SUPARNA DEEPAK, 202121054595, MAHATMA EDUCATION SOCIETY'S, PILLAI HOC COLLEGE OF ARTS, SCIENCE AND COMMERCE, "Novel Radar Absorbing Paint for Military Stealth Technology." 2021, Indian Patent office.
- Prajakta Harshad Kane, 202121054600, MAHATMA EDUCATION SOCIETY'S, Pillai College of Arts Commerce & Science, "Intelligent Computer Texter" 2021, Indian Patent office.
- Deepika Sharma, 202121054603, MAHATMA EDUCATION SOCIETY'S, Pillai College of Arts, Commerce & Science (Autonomous), Dr K M Vasudevan, "distance between user mobile phone and fitness band more than 5 mts (or user set distance) then alarm." 2021, Indian Patent office.
- Dr. Kavita Kathare, Dr. Aarti Sukheja, 202121054599, MAHATMA EDUCATION SOCIETY'S, Pillai College of Arts, Commerce & Science (Autonomous," show of the part which is harmed and must be supplanted in Washing machines" 2021, Indian Patent office.
- Dr. Divya Padmanabhan, Jweshvari Tupe, 202121054602,Mahatma Education Society's, Pillai College of Engineering "Natural Fiber Reinforced Composites for Filtration of Water" 2021, Indian Patent office.
- Dr. Reni Francis (Principal), Dharti Latke , Dr. Jaya Cherian, 202121054601, MES's Pillai College of Education and Research " Facial Emotion Recognition using Deep Learning and AI." 2021, Indian Patent office.

CHAPTER - 07

Real Time Concrete Strength Monitoring Using Machine Learning

Mr. Raju Narwade
Mahatma Education Society's, Pillai HOC College of Engineering and Technology
Mr. Aditya Jadhav
Mahatma Education Society's, Pillai HOC College of Engineering and Technology
Mr. Umakant Dinkar Butkar
Geh Research, India, Japan, USA, Singapore

ABSTRACT

Our Invention "In-situ IoT Cloud Based Concrete Maturity-strength Measurement Systems for Real Time Concrete strength Monitoring and Prediction" is to a Wireless Embedded Maturity-Strength Measurement System (WE-MSMS) is designed and developed using digital DS18B20 Temperature Sensor array probe, microcontroller and Zig-Bee network with application software for maturity-strength measurement by non-destructive approach to estimate the in-situ concrete compressive strength of structure. A portable Wireless Temperature Sensor node with Zig-Bee network can be placed at construction site in the fresh concrete initially. Four-Sensor Nodes GUI Java platform based Application software is designed and used in the main server. In each sensor node there are eight- digital DS18B20 Temperature Sensors. For the measurement DS18B20 Temperature Sensor array probe embed into the fresh concrete, each DS18B20 Temperature Sensor with unique code is recorded and maintain in the database at main server. Thereafter every 5-minute or according to time setting till 28-days, time-temperature updated information will recorded in the central database using Zig-Bee network and IoT cloud. The designed system is useful for civil engineer to real-time prediction of in-situ concrete compressive strength at anytime and anywhere with the help of wireless technologies and IoT cloud which decide in-place grading of concrete in concrete construction work. Wireless maturity-strength measuring systems (WE-MSMS) can be implemented on different construction sites and connected with IoT cloud for real-time prediction of in-situ concrete compressive strength.

BACKGROUND

In traditional cube test method for compressive strength measurement of concrete mix design (IS 10262:2009). This is time consuming, wastage of concrete material and laborious job. The invented system provides the following solutions:

- In-situ IoT cloud based concrete maturity-strength measurement systems for real time concrete strength monitoring and prediction.
- It can be monitored concrete compressive strength by non-destructive approach and concrete maturity method.
- It can be maintained central database of concrete maturity-strength record in the main server of construction sites from all placed Sensor Nodes for further analysis.
- It can generate real time concrete maturity measurement based on IoT cloud as per the civil engineer's requirement within a second.
- It can also predict the real time concrete compressive strength to prevent failure of the concrete structure.
- It can also operate in dual power modes either solar PV mode or electricity mode.

PRIOR ART SEARCH

Concrete is essential part of civil engineering for the construction of any architectural structure viz. building, parking structure, pavements, bridges, roads, runways, water storage dams, cement pipes and other purposes. During most of construction works it is found that there is a lack of understanding of exact concrete maturity-strength because one cannot measure maturity-strength in terms of any figure or by using any type of measuring instrument.

This was the inspiration for undertaking the project work to develop embedded system that can exactly measure concrete maturity-strength in-place on experimental basis which can directly help civil engineering department to decide the grading of concrete that is compressive strength of concrete used in the construction site.

There is a great importance of maturity-strength measurement for any type of concrete construction work. The most common and probable consequences of failure to maintain sufficient maturity-strength of any civil structure can cause dangerous results such as collapsing of entire structure or early visible cracks and leakages or bends in the concrete structure are seen due to premature construction work.

The concrete in-place maturity-strength monitoring information is important to determine structural strength for serving at early age. In current scenario, the wireless temperature sensor node based embedded technology helps for monitoring early ages of concrete in-place maturity-strength. The concrete maturity method uses the time-temperature factor history (TTF) of concrete to estimate compressive strength development during curing time period.

The designed and developed Embedded Wireless Maturity-Compressive Strength Measurement System (EW-MSMS) which is used for maturity-strength measurement consists of non-destructive method for estimating concrete compressive strength under investigation in the field during construction practice.

This EW-MSMS plays an important role to collect, process, store, transmit and display the data/information related to concrete compressive strength (CCS). The specialty of EW-MSMS is that the collected information can be accessed at any time and also in-place with the help of wireless communication network and IoT cloud.

Advantages Over the Existing Art

- Central system and real-time in-situ concrete-maturity measurement database control by non-destructive approach.
- Wireless Temperature Sensor Nodes, IoT cloud based, portable and handy concrete-maturity measurement system can directly help civil engineers in remote area.
- Time-temperature updated information recorded in the central system using Wireless Embedded Maturity-Strength Measurement System (WE-MSMS) by non-destructive approach and maturity method.
- Dual Mode of Power: Solar PV and electricity.
- Once it is activated, then it's on for 28-days for prediction of concrete compressive strength in real time.

- Time-temperature updated information will be recorded in Real time and more accurately.
- Maturity-Strength Reports of concrete construction using Nurse-Saul, Freiesleben Hansen and Pedersen and CEB-FIP mode are available within a click as per the civil engineers requirement.

OBJECTIVES OF THE INVENTION

- The objective of the invention is to provide a "Real Time Concrete strength Monitoring and Prediction" is to Embedded Maturity-Strength Measurement System (WE-MSMS) by non-destructive approach is designed and developed using digital DS18B20 Temperature Sensor array probe, microcontroller, Zig-Bee network and IoT cloud.
- The other objective of the invention is to provide a Zig-Bee network with application software for maturity-strength measurement by non-destructive approach to estimate the concrete compressive strength of structure. A portable Wireless Temperature Sensor node with Zig-Bee network can be placed in the fresh concrete initially at construction site.
- The other objective of the invention is to provide a thereafter every 5-minute or according to time setting till 28-days, time-temperature updated information will recorded in the database using Zig-Bee network and IoT cloud.
- The other objective of the invention is to provide an eight-DS18B20 Temperature Sensors in an each sensor node. Four-Sensor Nodes Java platform based Application software is designed and used in the construction site's main server.
- The other objective of the invention is to provide a designed system useful for civil engineer to real-time prediction of in-situ concrete compressive strength at any-time and any-where with the help of wireless technologies and IoT cloud and also provide the real-time grading of concrete in various type of construction work.
- The other objective of the invention is to provide wireless and Real Time - Temperature updated recording and monitoring, and the computation of concrete strength by Maturity (TTF), equivalent age and temperature adjusted age functions.
- The other objective of the invention is to implement wireless maturity-strength measuring systems (WE-MSMS) on different

construction sites and connected with IoT cloud for real-time prediction of in-situ concrete compressive strength.

- The other objective of the invention is to support ASTM C-1074 and IS 516:1959 Standards.

SUMMARY OF THE INVENTION

Existing System: CN865 Concrete Maturity Meter https://www.impact-test.co.uk/products/6581-6-channel-concrete-maturity-meter/	**Our system:** In-situ IoT cloud Based Concrete Maturity-strength Measurement Systems for Real Time Concrete strength Monitoring and Prediction
How it Works?	
The CN865 has 6 temperature channels and digitally displays the value of all channels.	It has 8 digital temperature probe with each Node
It can be connected directly to a PC or laptop via a USB interface for fast downloading of the readings.	Wireless connectivity via Zigbee network to a main construction site server.
Using the supplied Microsoft Excel™ templates, the Concrete Maturity is calculated from elapsed time versus temperature and is quoted as a "Maturity number".	Concrete Maturity calculation and prediction of concrete compressive strength using Excel.
The Maturity number can be used to determine when concrete has cured sufficiently for formwork to be removed.	Maturity and Compressive strength number can be used to remove the formwork.
In order to determine concrete maturity, sacrificial temperature measurement cables are 'cast into' the concrete and cut off at the surface when the concrete has cured.	Digital one-wire DS18B20 used to determine concrete maturity.
This Concrete Maturity Meter is a compact unit, making it ideal for site work in all conditions and climates.	Compact and wireless its ideal for site construction work in all conditions and climates.
The maturity meter is simple to use, finishing the job in the shortest possible time.	Wireless and in-situ
Features & benefits	
Logger for 6 x K or T type probes	32- Digital one-wire DS18B20 probe
Type K extension leads	8- Temperature sensors in Each node.

Software, suitable for Windows 2000, XP and Vista	Software Windows-7 and Windows-10
USB connecting cable	Wireless connectivity and IoT cloud
Our standard thermocouple wire & plugs may be used, for example TH088 & TH095.	Digital one-wire DS18B20 probe used.
Benefits with the implementation of Concrete Maturity Measurement Systems.	
Concrete Maturity Meter is a compact unit	Handy and compact
Concrete maturity measurement	in-situ concrete maturity and concrete strength measurements based on time-temperature history and age.
Support ASTM C-1074 & ASTM C-918	Support ASTM C-1074 and IS 516:1959 Standards.
Construction personnel with minimal training can easily implement monitoring function.	Same

BRIEF DESCRIPTION OF THE DIAGRAM

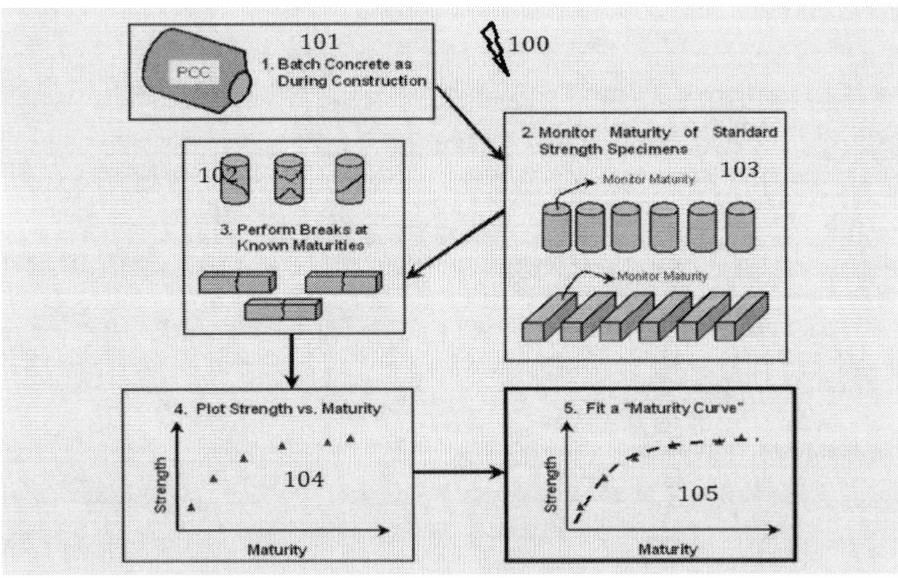

Fig.1: In-situ IoT cloud Based Concrete Maturity-Strength Measurement Systems flow chart (Prior Art Search)

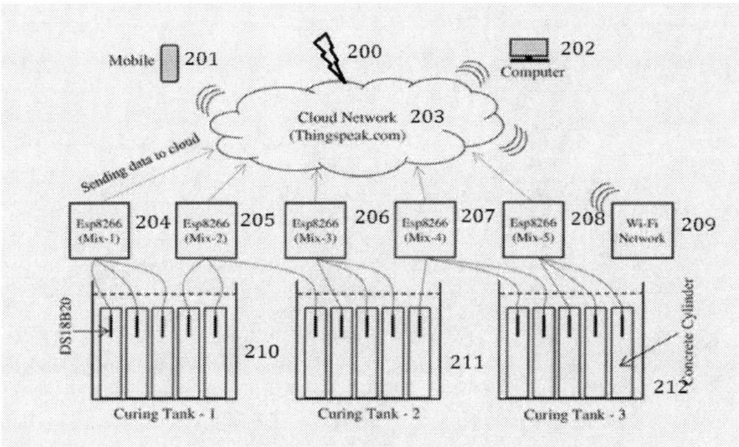

Fig. 2: In-situ IoT cloud Based Concrete Maturity-Strength Measurement Systems for Real Time Concrete strength Monitoring and Prediction (Prior Art Search)

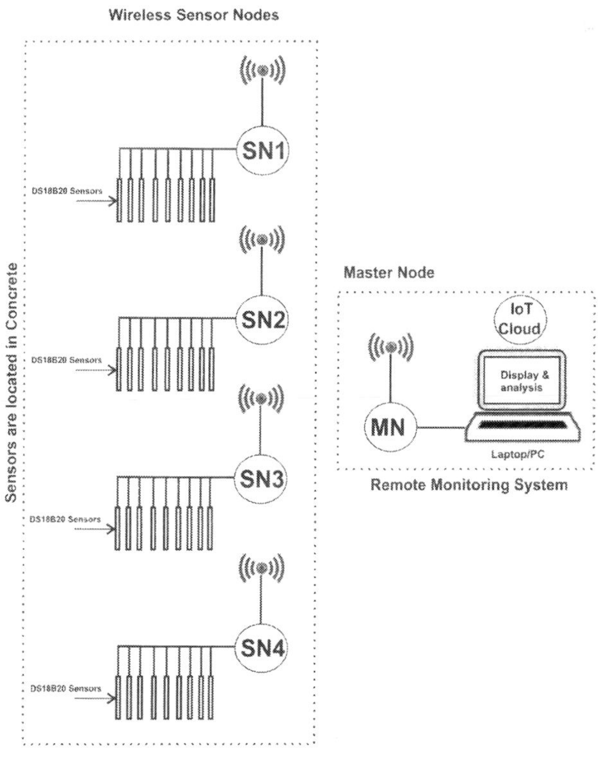

Fig. 3: Basic block of wireless sensor network for EW-MSMS

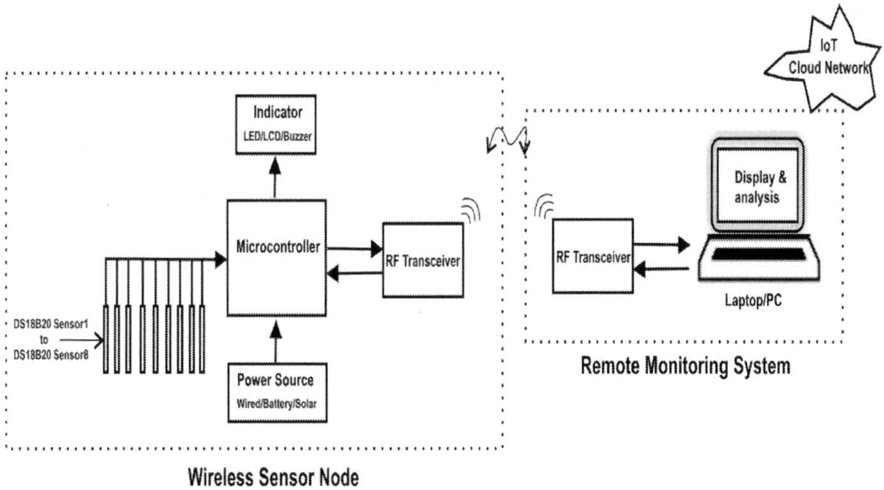

Fig. 4: Basic block of wireless sensor network for EW-MSMS

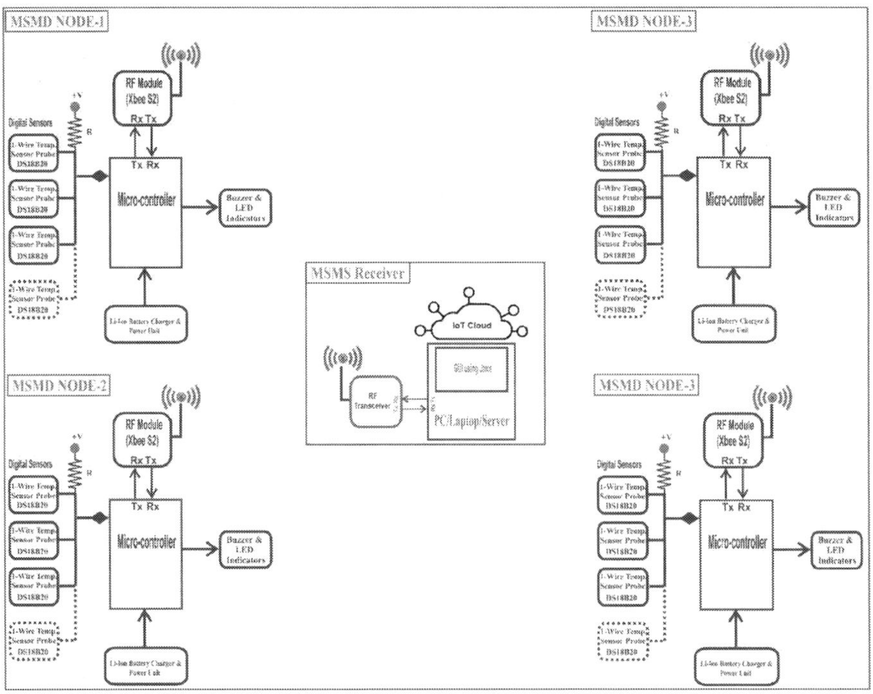

Fig., 5: Functional Block diagram of Wireless Embedded Maturity-Strength Measuring System

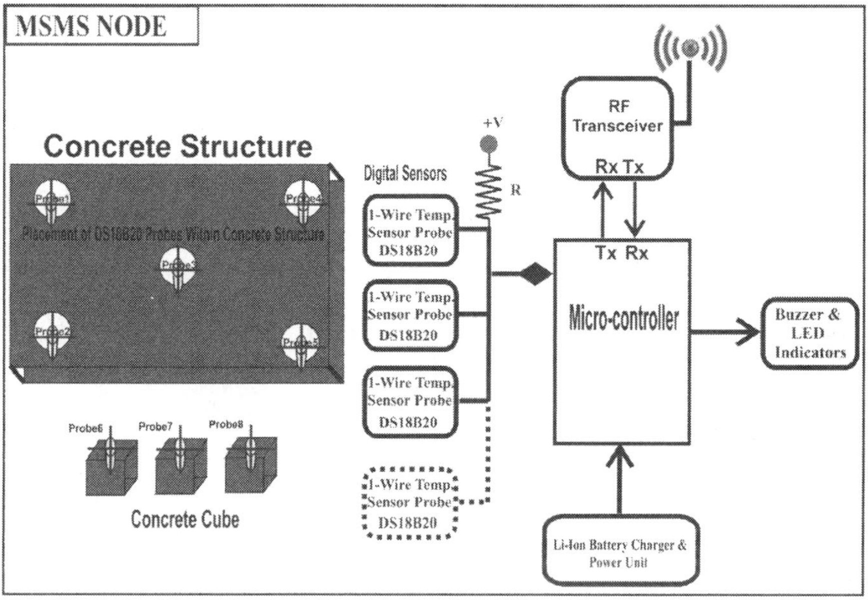

Fig. 6: Block Diagram of EW-MSMS Node.

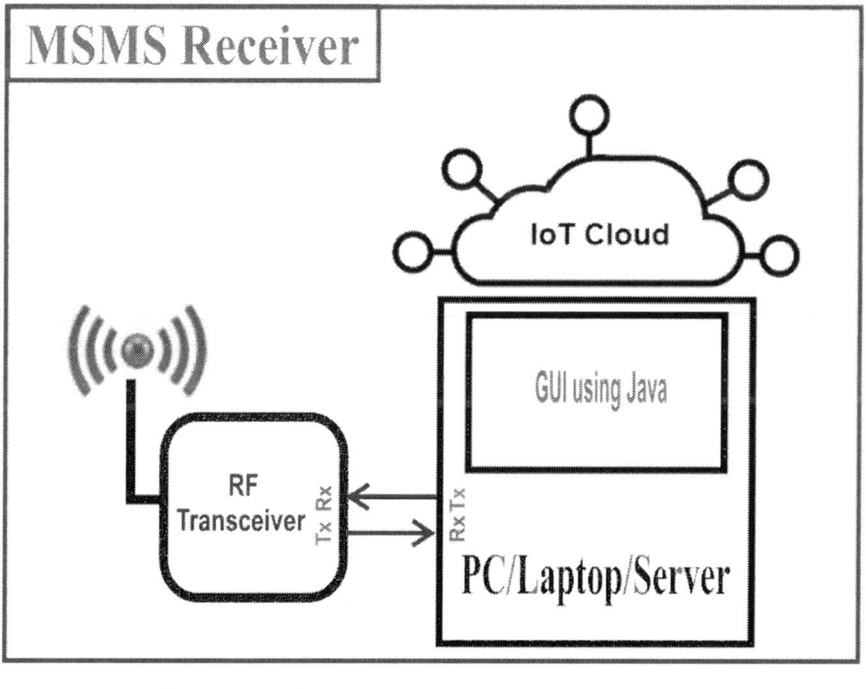

Fig. 7: Block Diagram of EW-MSMS Receiver

Fig. 8: Complete EW-MSMS Setup

DESCRIPTION OF THE INVENTION

Wireless Sensor Network (WSN) for Maturity-Strength Measuring System:

- The designed wireless sensor network (WSN) is shown in figure 3. The network consists of sensor nodes (SNs) and master node (MN). The WSN consist of four wireless sensor nodes, namely, SN1, SN2, SN3 and SN4.
- These SNs are located at different places in the construction field and all temperature sensors are embed in the concrete and sending temperature information to the master node (MN) for the estimation of in-place maturity-strength of the concrete. Then send the same information on cloud for access and further analysis.
- The received data by SN1, SN2, SN3 and SN4 are displayed on the laptop to allow the user to monitor the concrete maturity and predict compressive strength and the same on cloud. MN is connected to the laptop, thus requiring no battery power source. The SN1, SN2, SN3 and SN4 are away from MN and their energy

is derived from their individual power supplies. This means that SN1, SN2, SN3 and SN4 can be powered by the main AC source or battery or solar power source.

Wireless Sensor Node (SN) for Maturity-Strength Measuring System:

Figure 4: show the basic block of wireless sensor node (SN) for Maturity-strength measuring system. It consists of wireless sensor node and remote monitoring system.

Figure 4: Basic block of wireless sensor network for EW-MSMS

Each wireless sensor node is equipped with maximum eight numbers of 1-wire digital temperature sensors DS18B20, an ATMega 328P Microcontroller, Xbee transmitter, LED and buzzer indicator, and LCD display unit (optional). The entire sensor node (SN) is powered by the use of Li-ion battery backup power supply with charging circuitry arrangement. The sensor nodes (SNs) comprises of temperature sensors for measurement of temperature parameter of concrete structure and also concrete cube samples.

The remote monitoring system consist of Zig-Bee receiver in connection with personal computer or laptop or desktop or even to the raspberry Pi. The Graphical User interface is designed using JAVA language platform for monitoring and the measurement of thermal parameter and analysis of early age (fresh) concrete.

This arrangement is done for online real-time displaying temperature parameter recorded by each sensor embed in concrete cube which is under measurement. Hence the same arrangement is applied for finding the in-situ maturity-strength of concrete structure in field.

Wireless Embedded Maturity-Strength Measuring System (EW-MSMS)

- The entire functional block diagram of wireless embedded Maturity-strength measuring system is as shown in figure 5. It consists of four sensor nodes (SNs) and remote monitoring system for data acquisition and thermal analysis of early age (fresh) concrete.
- The entire Wireless Embedded Maturity-Strength Measuring System (EW-MSMS) is divided into five functional sections as follows,
- Sensing temperature data from concrete structure /concrete mix design cubes by using multiple temperature sensors.

- Data acquisition from each digital temperature sensor and to process on received data.
- Data transmission using Zig-Bee wireless communication devices
- Data reception using Zig-Bee wireless communication device.
- Data analysis of received data by personal computer/Laptop.
- The detail description of each functional block of Wireless Embedded Maturity-Strength Measuring System (EW-MSMS) as follows:

EW-MSMS Transmitter Node

Fig.6: The EW-MSMS measurement system's each data transmitting node consists of multiple numbers of 1-wire digital temperature sensor embed in concrete cube that are interfaced with an AT Mega 328PU eight bit microcontroller, a Xebec transmitter device for wireless data transmission, a Li-Ion battery unit for power backup with charging facility and LED and Buzzer indicator. In the present system there are 4 – nodes.

EW-MSMS Receiver Node

The EW-MSMS measurement system's data receiver section as shown in figure 7. consists of Xbee receiver connected to the personal computer terminal as shown in figure8. It is equipped with data acquisition and GUI software and data analysis software for the prediction of maturity strength of concrete cube under measurement.

Hardware Setup of Wireless Embedded MSMS

The complete hardware setup of Wireless embedded Maturity-Strength measurement system consisting Laptop, four Sensor Nodes (SNs), and remote receiver master node (MN) with IoT cloud is shown in figure 8.

Present Methods and Mathematical Models

Concrete is one of the most widely used in mixture form for any form of architectural structures, pavements, bridges, roads, runways, parking structures, dam and pipes etc. During initial state of concrete, in first few hours of consolidation if proper care is not taken, one can easily observe early cracks in solidified structure or leakages in structure and even sometimes entire structure can collapse that may turn into disastrous situation because of premature construction work.

Therefore, it is highly and most essential to measure and monitor the maturity strength of the concrete using in-place non-destructive technique is the only scientific way to prevent failure of the concrete structure. This is the main aspect in concrete structure development, the main emphasis is on strength measurement of the concrete structure.

There are also two major methods used for measuring the concrete strength, viz. destructive and non-destructive method. Destructive maturity strength measuring methods are time consuming as well as costly and are not suitable for the early age of concrete strength monitoring. These methods are not available at construction sites. Mostly, the concrete maturity was calculated using the Nurse-Saul and Arrhenius mathematical model/equations.

The present non-destructive methods are suitable for early ages of concrete strength monitoring. These are based on Thermocouples, Dallas Semiconductor iButton®, Nomadics, intelliRock™ and Identic Solutions i-Q Tag devices.

But there are some drawbacks of these systems such as bulky in size, direct wired measurement on site, limited channels, processing software limitations, non-continuous monitoring and costly.

The presented EW-MSMS is fully wireless technique based on the Nurse-Saul and Arrhenius mathematical model/equations. It minimizes drawbacks mentioned above. The key features of presented work are mentioned below.

- Multichannel facilities
- Wireless measurement on the site at anywhere and anytime
- Compact in size and Portable
- Software support (Java based GUI application)
- 24X7 Continuous monitoring using wireless techniques and IoT cloud
- Compact system setup
- Low Cost

Destructive method and non-destructive methods are used for measurement of concrete maturity. Destructive method was used for the actual physical measurement of strength of concrete by applying appropriate mechanical force on specimen structure. The performed during actual curing process of samples, it results in achieving compressive strength for sample under various environmental conditions.

The results obtained from non-destructive EW-MSMS correlated in concrete strength measurement process. The measured parameters from the distinct nodes are used for analysis and also prediction of Concrete Compressive Strength (CCS) using Raspberry Pi -3 based minicomputer or computer.

The compressive strength of concrete is measured at two places simultaneously i.e. in laboratory and on construction site. It is measured at ages of 1, 3, 7, 14 and 28 days in laboratory by using cube test and this data is stored in Micro SD card of the Raspberry Pi-3 minicomputer or computer. At the same time, it is measure on construction site and transfer and available the information on web using IoT cloud. The prediction of the compressive strength of the structure is represented in graphical format i.e. deviation of concrete compressive strength verses age using GUI designed with JAVA based software.

The wireless system is designed and implemented for monitoring, analysis and also prediction of compressive strength of the concrete which is useful to prevent architectural structure failure. In sensor nodes are designed to measure temperature development in early age concrete structure using ATmega328PU microcontroller based Arduino platform which transmits the sensed parameters in real time to the Raspberry Pi-3 minicomputer at receiving end.

The IoT cloud based Raspberry Pi-3 or minicomputer acts as receiver used for data analysis and maturity strength prediction purpose. Received data from Arduino platform transmitting nodes was also sent on data cloud using Internet platform anywhere using internet connectivity.

The designed system has four - wireless nodes. Each wireless node consists of eight (08) DS18B20 temperature sensors to measure temperature inside the concrete at construction site. Each wireless node transmits sensed temperature parameter to the base station using Zig-Bee wireless communication.

The base station consists of combination of internet, Zig-Bee and Raspberry Pi-3 minicomputer or computer. The received data is used for analysis and prediction of concrete strength using JAVA based Graphical User Interface (GUI).

Material and Method

The present system is divided into two parts; Transmitter Section using Zig-Bee and Arduino Platform and Receiver Section using Zig-

Bee and Raspberry Pi Platform or IoT cloud based computer. The transmitter section consists of wireless nodes using Zig-Bee. In four nodes are used. Each node consists of following components.

Transmitter Section

Temperature Sensors:

The DS18B20 is temperature sensing module which is used for temperature measurement. It is a small, 1-wire sensor digital sensing module with built in 12bit ADC. It gives temperature output in terms of degree Celsius or degree Fahrenheit. In each node, eight DS18B20 temperature sensors are immersed in concrete cube / structure.

Microcontroller:

The Atmega 328PU microcontroller is used as a part of sensor node. It is 28-pin, a single-chip microcontroller manufactured by Atmel. The output of each temperature sensor is connected to ATmega 328PU microcontroller using 1-wire interface. The ATmega 328PU microcontroller is easy to interface with other system using Arduino platform.

Zig-Bee Module:

The Xbee S2C is used for wireless data transmission. The measured temperature parameters in Celsius collected by ATmega328PU microcontroller are sent to the base station using Zig-Bee module. The Zig-Bee is configured as End device. The configuration of Xbee S2C has done through the X-CTU software by Digi International.

Receiver Section

The receiver section also called as base station. It consists of Raspberry Pi-3 microcomputer or computer and data receiving Zig-Bee module. This section consists of following parts.

Zig-Bee Module:

It is at receiver section. This Zig-Bee is configured as coordinator. The data sent by transmitter section is collected at base station using Zigbee module. The configuration of Zig-Bee S2 has done through the X-CTU software by Digi International

Raspberry Pi and computer:

Raspberry Pi is 64-bit minicomputer with Broadcom BCM2837 and Quad Core Cortex A53. It work as mini computing device with Raspbian operating system. The data is stored in Raspberry Pi or computer for analysis.

Graphical User Interface (GUI):

The JAVA based GUI is developed for data analysis. The graphical analysis is done for temperature-time history. By using this graphical information, concrete strength can be predicated to avoid any structural failure due to premature work.

Wireless communication

The data stored in Raspberry Pi or computer can be sent on cloud using Internet (IoT). Also this data and graphical information can be shared on email to user.

Embedded Wireless-Maturity-Strength Measurement System (EM-MSMS)

EW-MSMS is a real-time, wireless communication based non-destructive way for continuous monitoring and analyzing for the in-place measurement of concrete compressive strength (CCS) at early ages. Destructive and Non-destructive tests are conducted for estimating compressive strength of concrete in laboratory (cube) and on the field (during construction).

The difference between two types of tests results was found in the range of 10% to 20% at approx. Thus Embedded Wireless-Maturity-Strength Measurement System (EW-MSMS) methodology has more potential to be used in-place compressive strength accurately for better results in concrete maturity under measurement.

Maturity-Strength Measuring System Hardware Design

The hardware details of the present EW-MSMS system which is useful for prediction of concrete compressive strength. Hardware design part includes Temperature sensor node (SN), wireless Transmitter unit and wireless Receiver Unit.

The wireless sensor network (WSN), wireless sensor node (SN) and various components used in WSN. It also covers details, features and application of selected devices such temperature sensors (DS18B20), microcontroller (ATMega328P) & Zig-Bee (Xbee-S2C) in the present system.

Maturity-Strength Measuring System Software

The software details of present works. Software tools includes WSN configuration with X-Bee S2c using X-CTU software platform, Sensor Node (SN) coding with ATMega328P microcontroller using Arduino Uno software platform and graphical user interface programming (GUI) using JAVA software platform. It also cover the development of data analysis forms for the presentation of measured data is discussed in detail.

Field Results

The about experimental work carried out on comparison results of concrete mix design cube test of M15, M20, & M25 samples using universal testing machine and discuss the results obtained from the above mentioned tests using wireless maturity-strength measuring system (WE-MSMS).

On basis of data collection made by present system the same utilized for measurement of compressive strength function or model suggested by Nurse-Saul Maturity Method, F. Hansen and Pedersen function and CEB-FIP model. It also discloses.

OUTCOME

Our Invention "In-situ IoT Cloud Based Concrete Maturity-strength Measurement Systems for Real Time Concrete strength Monitoring and Prediction" is to a Wireless Embedded Maturity-Strength Measurement System (WE-MSMS) is designed and developed using digital DS18B20 Temperature Sensor array probe, microcontroller and Zig-Bee network with application software for maturity-strength measurement by non-destructive approach to estimate the in-situ concrete compressive strength of structure. A portable Wireless Temperature Sensor node with Zig-Bee network can be placed at construction site in the fresh concrete initially. Four-Sensor Nodes GUI Java platform based Application software is designed and used in the main server. In each sensor node there are eight- digital DS18B20 Temperature Sensors. For the measurement DS18B20 Temperature Sensor array probe embed into the fresh concrete, each DS18B20 Temperature Sensor with unique code is recorded and maintain in the database at main server. Thereafter every 5-minute or according to time setting till 28-days, time-temperature updated information will recorded in the central database using Zig-Bee network and IoT cloud. The designed system is useful for civil engineer to real-time prediction of in-situ concrete compressive strength at anytime and anywhere with the help of wireless technologies and IoT cloud which decide in-place grading of concrete in concrete construction work. Wireless maturity-strength measuring systems (WE-MSMS) can be implemented on different construction sites and connected with IoT cloud for real-time prediction of in-situ concrete compressive strength.

- According to claim1# the invention is to an "In-situ IoT cloud Based Concrete Maturity-strength Measurement Systems for Real Time Concrete strength Monitoring and Prediction" is to a Wireless Embedded Maturity-Strength Measurement System (WE-MSMS) is designed and developed using digital DS18B20 Temperature Sensor array probe and microcontroller and also Zig-Bee network with application software for maturity-strength measurement by non-destructive approach to estimate the concrete compressive strength of structure.
- According to claim1,2# the invention is to an portable Wireless Temperature Sensor node with Zig-Bee network can be placed at construction site in the fresh concrete initially. Four-Sensor Nodes GUI Java platform based Application software is designed and used in the main server.
- According to claim1,2,3# the invention is to an portable Wireless Temperature sensor node there are eight- DS18B20 Temperature Sensors in array probe.
- According to claim1,2,3,4# the invention is to the DS18B20 Temperature Sensor array probe embed into the fresh concrete, each DS18B20 Temperature Sensor with unique code is recorded and maintain in the database at main server.
- According to claim1,2,3,4,5# the invention is to an 5-minute or according to time setting till 28-days, time-temperature updated information will recorded in the database using Zig-Bee network and IoT cloud.
- According to claim1,2,4,5,6# the invention is to an designed system is useful for civil engineer to real-time prediction of in-situ concrete compressive strength at anytime and anywhere with the help of wireless technologies and IoT cloud and decide the grading of concrete in construction work .
- According to claim1,2,4,5,6,7# the invention is to an portable Wireless Temperature Sensor node can be powered by the main AC source or battery or solar power source for continuous system functioning.
- According to claim1,2,4,5,6,7,8 # the invention is to an wireless maturity-strength measuring systems implemented on different construction sites and connected with IoT cloud for real-time prediction of in-situ concrete compressive strength.

CITATIONS

- Dr. Sanjay B. Munavalli, 202121056484, Pillai College of Arts, Commerce & Science (Autonomous)
- Dr K M Vasudevan PILLAI Campus, Sector 16, New Panvel, NAVI Mumbai -410206, MH, India, "Intelligent Digital Water Purifier (IDWP) with Filter Quality Indicator (FQI) and Water Consumption Indicator (WCI)" 2021, Indian Patent office.
- Dr. Abida Khan, 202121056482 Vasudevan PILLAI Campus "Department of B.Com. Accounting & Finance, Pillai College of Arts, Commerce & Science (Autonomous Dr K M, Sector 16, New Panvel, NAVI Mumbai -410206, MH, India, " Colouring Crayons/Pencils which kids can use to scribble only on surface of notebook/pages and not on wall/floor/furniture." 2021, Indian Patent office.
- Dr. Kiran Sagar Deshmukh, 202121056485, Department of B. Com, Pillai College of Arts Commerce and Science (Autonomous), Dr K M Vasudevan PILLAI Campus, Sector 16, New Panvel, NAVI Mumbai -410206, MH, India ," Intelligent Electric Vehicle Battery Supper Fast Stations" 2021, Indian Patent office.
- Dr. C K Prashant, Suparna Deepak, 202121056488," Department of Biotechnology, Pillai college of Arts, Commerce & Science (Autonomous), Dr K M Vasudevan PILLAI Campus, Sector 16, New Panvel, NAVI Mumbai -410206, MH, India,
- "Method and Process for Targeting Intra-Cellular Bacterial Infections Using Nanoparticles. "2021, Indian Patent office
- Ms. Remya Madan Gopal, 202121056486, MAHATMA EDUCATION SOCIETY'S Pillai HOC College of Arts, Science and Commerce, Pillai HOCL Educational Campus, Rasayani, Taluka Khalapur, Dist, Raigad, Maharashtra 410207, "Cancer Therapeutics Using Nanoparticles Based on Warburg Effect." 2021, Indian Patent office.
- MAHATMA EDUCATION SOCIETY'S ,202121056483, MAHATMA EDUCATION SOCIETY'S PILLAI HOC COLLEGE OF ENGINEERING AND TECHNOLOGY, "Experimental Studies on the Use of Waste Rubber Tyre Crumb and Bamboo in Tremix Concrete Flooring" 2021, Indian Patent office.
- Suprita Manohar Rao, Sadhana Shivkumar Thevar, 202121057477, Mahatma Education Society's Pillai College of Arts, Commerce & Science (Autonomous), Dr K M Vasudeva PILLAI Campus, Sector 16, New Panvel, NAVI Mumbai -410206,

MH, India, "Intelligent Methods and Process to Rc- Cycle the Waste Plastic" 2021, Indian Patent office.

- Ms. Meenakshi Johri, Ms. Bindu Rajaguru, Mr. Gopakumar Pillai, 202121057473, MAHATMA EDUCATION SOCIETY'S PILLAI COLLEGE OF ARTS, COMMERCE & SCIENCE (AUTONOMOUS), " Nutraceutical Product Combinig Wheat Grass and Fenugreek" 2021, Indian Patent office.

CHAPTER - 08

Intelligent Drainage Using Machine Learning

Pragnesh Narendra Shah

Mahatma Education Society's, Pillai College of Engineering, Dr K M Vasudevan PILLAI Campus, Sector 16, New Panvel, NAVI Mumbai -410206, MH, India.

Prof (Dr.) B.K. Sarkar

MES's, Pillai HOC College of Engineering & Technology, Rasayani Taluka Panvel, Dist, Navi Mumbai, Maharashtra 410207, India

Dr. Vinay Rishiwal

Professor, MJP Rohilkhand University, Bareilly-243006, UP, India.

ABSTRACT

Our invention "Intelligent Drainage using Machine Learning" is a Manhole maintenance by human control is very difficult because of the environmental issues and also it is difficult to go inside each manhole for inspecting its state. This project aims at developing a Smart Drainage System using Internet of Things (IoT). In this system there are sensors placed inside of the manhole which detects and transfers appropriate information about water levels, slit deposit and other factors to the managing station (Municipal Corporation) using the Android application? So that they can take precautionary action to solve the issue considering public safety. This project also aims to solve the problem of manhole lid displacement by using the tilt sensor.

BACKGROUND OF THE INVENTION

The Internet of Things (IoT) consists of real life objects, communication devices attached to sensor networks in order to provide communication and automated actions between real world and information world.

Though, computers were able to access data from objects and devices without human intervention, IoT came into existence with an aim to overcome the limiting factors of human entered data, and to achieve cost, accuracy and generality factors. Sensor Network is a key enabler for IoT paradigm.

Drainage maintenance is necessary to keep the city clean, safe and healthy. To overcome the problems faced by traditional drainage system most of the cities adopted underground drainage system. But also if underground drainage gets blocked, it will create many problems such as traffic jam, the environment becomes dirty, and if manhole lid is not closed properly there is a chance of occurrence of accidents and also people may fall inside the manhole.

To overcome all these problems it is necessary to adopt a Smart Drainage system for transmitting and storing current states of information detected by sensors set inside the manhole to a managing station so that the necessary actions can be taken on time and accidents could be avoided.

OBJECTIVES OF THE INVENTION

The objective of the invention is to a develop a IoT based Smart Drainage model that is able to:

- The other objective of the invention is to a Detect the water level inside the manhole and inform about the same to Municipal Corporation
- The other objective of the invention is to a Detect water flow inside the manhole and inform about the same to the Municipal Corporation
- The other objective of the invention is to a Detect the manhole lid displacement using the tilt sensor
- The other objective of the invention is to a Store the values into the android application.

SUMMARY OF THE INVENTION

Proposed System

As per the scope of the project our main aim is to develop an IoT based Smart Drainage System that will detect the problems occurred inside a manhole by the respected sensors placed inside and take actions accordingly and also automatically solve the problem of water clogging if occurred.

Initially all the sensors placed inside a manhole will be connected to Node MCU microcontroller. These sensors will sense water level, weight of garbage, Lid displacement send the alert to managing station (Municipal Corporation) with the help of Android Application and the managing station can then take the necessary action.

For receiving this alert the managing station needs to have the mobile application of Smart Drainage System. Also using this application, the managing system will be able to do real-time monitoring of water level, water flow, lid displacement and slit deposit.

The system will be connected to a Firebase (Database) where all the real time information received by the sensors will be stored. Therefore the managing station can check the live status of the manhole.

Also the android application will be storing and displaying the live data which is fetched from the sensors and transferred to firebase database.

IMPLEMENTATION

Android Studio

Android Studio is the official integrated development environment (IDE) for Google's Android operating system, built on Jet Brains' IntelliJ IDEA software and designed specifically for Android development. It is available for download on Windows, macOS and Linux based operating systems. It is a replacement for the Eclipse Android Development Tools (ADT) as the primary IDE for native Android application development. Android Studio was announced on May 16, 2013 at the Google I/O conference.

Circuit Design

Proteus 8: The Proteus Design Suite is a proprietary software tool suite used primarily for electronic design automation. The below figure represents the Schematic diagram of the Printed Circuit Board (PCB) for the Smart Drainage System which is used to show all the components and their connection in the circuit.

Arduino IDE: The Arduino integrated development environment (IDE) is a cross-platform application (for Windows, macOS, Linux) that is written in the programming language Java. It originated from the IDE for the languages Processing and Wiring. The below attached figure shows the coding screen of the Arduino IDE where all the necessary code is uploaded for Node MCU and all the sensors to make the system working.

BRIEF DESCRIPTION OF THE DIAGRAM

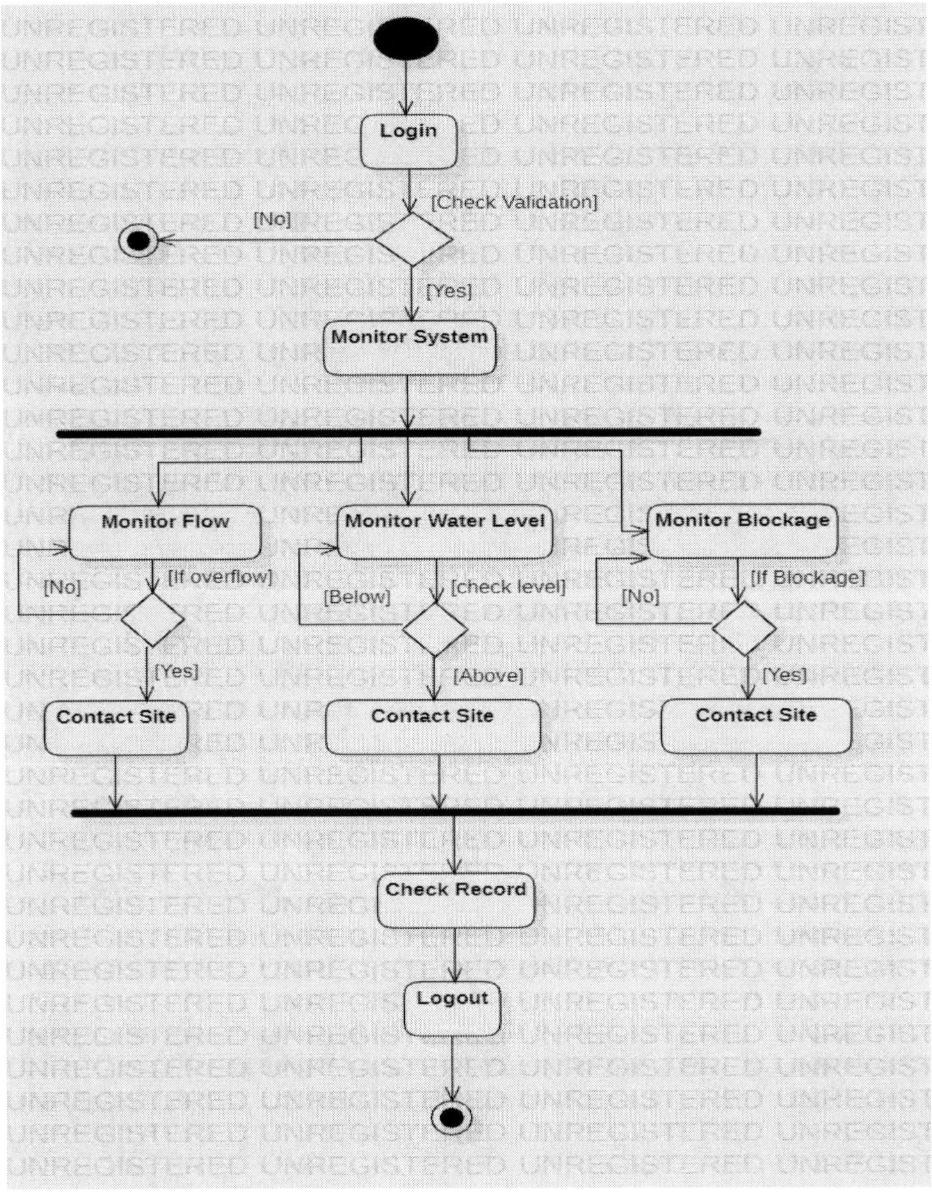

Fig. 1: Activity Diagram

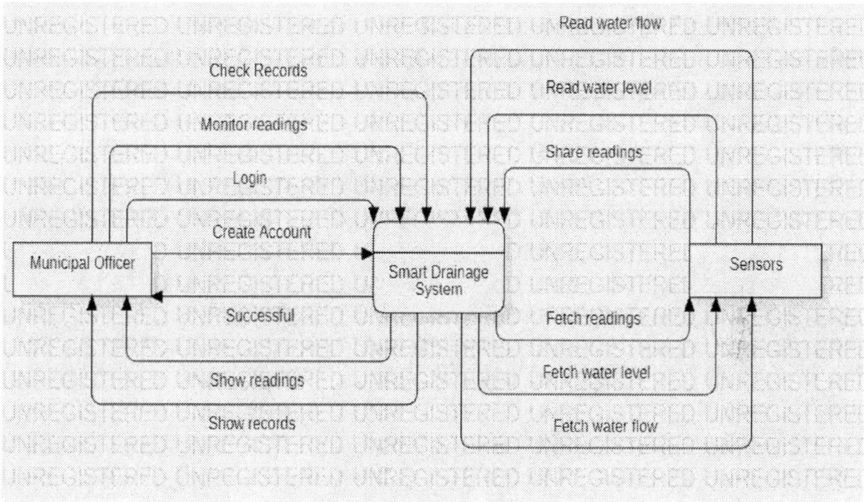

Fig. 2: Dataflow Diagram

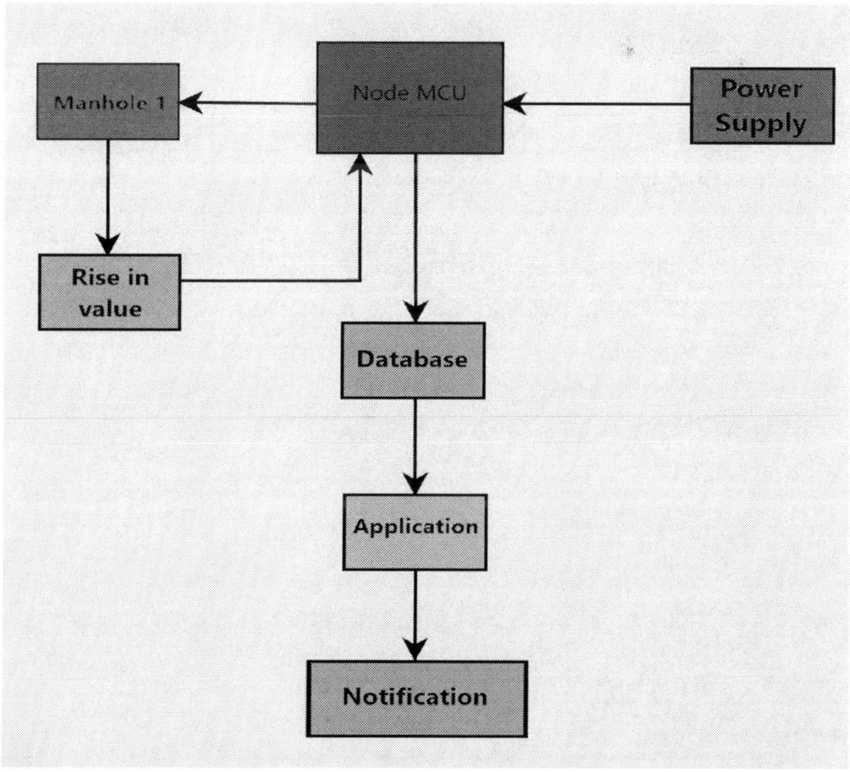

Fig. 3: System Architecture

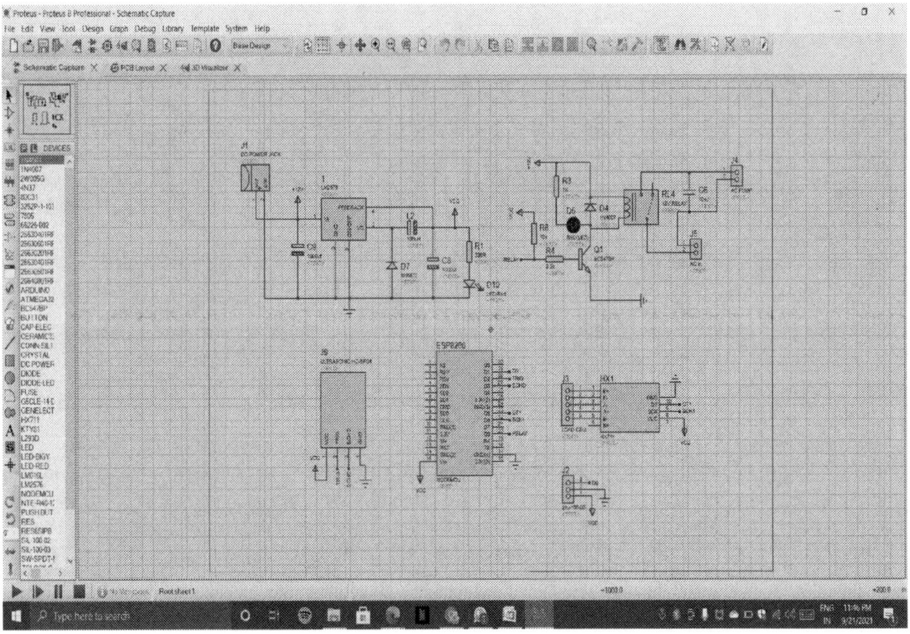

Fig.4: Schematic Diagram

DESCRIPTION OF THE INVENTION

Hardware and Software Requirements

Developing an IoT based system has some requirements especially when it is going to perform a particular task automatically. Following are the hardware and software requirements:

Hardware Requirements:

- Node MCU
- Tilt sensor
- Load sensor
- Ultrasonic sensor
- Power Supply
- 12 volt SMPS or an adaptor

Software Requirements:

- Firebase
- Android Studio

- Arduino IDE
- Proteus 8
- Description

Internet of Things (IoT)

The Internet of Things (IoT) consists of real life objects, communication devices attached to sensor networks in order to provide communication and automated actions between real world and information world. Though, computers were able to access data from objects and devices without human intervention, IoT came into existence with an aim to overcome the limiting factors of human entered data, and to achieve cost, accuracy and generality factors. Sensor Network is a key enabler for IoT paradigm.

Wireless Sensor Network (WSN)

Wireless Sensor Network (WSN) is the technology in which the data collected from the field of interest is transmitted through wireless link. WSN can be used in various fields such as monitoring, wireless measurements, controlling, etc. The Wireless Sensor Network (WSN) is integrated with the Internet of Things (IoT) to achieve the above objective.

Node MCU

The Node MCU is an open-source microcontroller board based on the Microchip ATmega328P microcontroller and developed by Arduino. cc. The board is equipped with sets of digital and analog input/output (I/O) pins that may be interfaced to various expansion boards (shields) and other circuits.

The board has 14 digital I/O pins (six capable of PWM output), 6 analog I/O pins, and is programmable with the Arduino IDE (Integrated Development Environment), via a type B USB cable. It can be powered by the USB cable or by an external 9-volt battery, though it accepts voltages between 7 and 20 volts.

Result

Firebase is a mobile and web application development platform developed by Firebase, Inc. in 2011, then acquired by Google in 2014. Firebase's first product was the Firebase Real-time Database, an API that synchronizes application data across iOS, Android, and Web devices, and stores it on Firebase's cloud. The below figure shows the real time data which is collected by all the sensors which are connected to the NodeMCU. These data is constantly reflected into the firebase database.

Conclusion

Thus, we conclude that we have partially accomplished our goal of developing a Smart Drainage system that detects the water level, detects tilt in manhole lid and slit deposit inside the manhole. All the data collected by the sensors gets stored into a Firebase (Database). The data is also reflected and stored into the Smart drainage android application used by managing station.

OUTCOME

Our invention “Smart Drainage System using IOT” is a Manhole maintenance by human control is very difficult because of the environmental issues and also it is difficult to go inside each manhole for inspecting its state. This project aims at developing a Smart Drainage System using Internet of Things (IoT). In this system there are sensors placed inside of the manhole which detects and transfers appropriate information about water levels, slit deposit and other factors to the managing station (Municipal Corporation) using the Android application? So that they can take precautionary action to solve the issue considering public safety. This project also aims to solve the problem of manhole lid displacement by using the tilt sensor.

- According to claim1# the invention is to a “Smart Drainage System using IOT” is a Manhole maintenance by human control is very difficult because of the environmental issues and also it is difficult to go inside each manhole for inspecting its state.
- According to claim1, 2, 3# the invention is to a project aims at developing a Smart Drainage System using Internet of Things (IoT). In this system there are sensors placed inside of the manhole which detects and transfers appropriate information about water levels, slit deposit and other factors to the managing station (Municipal Corporation) using the Android application?
- According to claim1, 2, 3# the invention is to that they can take precautionary action to solve the issue considering public safety. This project also aims to solve the problem of manhole lid displacement by using the tilt sensor.

CITATIONS

- Miss. Madhulika Sinha, Adith Selvaraj, Vishnu Panicker, Amol Niladhe, Prathamesh Pawar, 202121062302, MAHATMA EDUCATION SOCIETY'S, PILLAI HOC COLLEGE OF ENGINEERING &TECHNOLOGY RASAYANI, Paver blocks of Concrete with Recycled Aggregates and Partial Replacement of Cement with Alccofine", 2022, Indian Patent office.
- NAIR ROHAN UNNIKRISHNAN, 202221000050, "प्राबल्यम-ASCENDING" MAHATMA EDUCATION SOCIETY'S, PILLAI COLLEGE OF ENGINEERING, 2022, Indian Patent office.
- Ms. CHAITALI R. KULKARNI, Mr. RAJU NARWADE, Mr. KARTHIK NAGARAJAN, DR. G R PATIL, 202221000049, "Smart Eco-friendly E-waste Divider Block (SEED)" MAHATMA EDUCATION SOCIETY'S, PILLAI HOC COLLEGE OF ENGINEERING &TECHNOLOGY RASAYANI, 2022, Indian Patent office.
- Dr. Daphne Pillai, 2022210010567, Director, Pillai College of Arts, Commerce and Science," New methods of education to meet post-Covid requirements" MES's Pillai College of Arts, Commerce and Science, Dr. K. M. Vasudevan Pillai Campus, Sector 16, New Panvel, 401206 Navi Mumbai, India, 2022, Indian Patent office
- 202021106920, Granted, Dr. Sharda Sharma, Mathoscope, MAHATMA EDUCATION SOCIETY'S, Dr. Pillai Global Academy, Plot no.1, RSC-48, Gorai 2, Borivali (West), Mumbai, 400092.

CHAPTER - 09

Light Weight High Strength Material Using Hybrid Ultrasonic and Machine Learning

Prof (Dr.) Shilpa Pankaj Kewate

Mahatma Education Society's, Pillai Hoc College of Engineering &Technology, Panvel, Dist, Navi Mumbai, Maharashtra 410207, India.

Rohit Devappa Gawade

Mahatma Education Society's, Pillai Hoc College of Engineering &Technology, Panvel, Dist, Navi Mumbai, Maharashtra 410207, India.

Prakashdev Indradeo Singh

Mahatma Education Society's, Pillai Hoc College of Engineering &Technology, Panvel, Dist, Navi Mumbai, Maharashtra 410207, India

ABSTRACT

Pulse electrochemical machining process is hybridized with ultrasonic machining process and a new ultrasonic assisted pulse electrochemical machining setup is developed. The ultrasonic stack assembly capable of vibrating the electrochemical machining tool at 28-kHz frequency is successfully developed. Standard 230V, AC and 50Hz frequency electrical input is converted into 28,000Hz, 300V electrical signal with the help of an ultrasonic generator. The main working elements of the ultrasonic assisted pulse electrochemical machining (USAPECM) setup are cathode tool, tool feed mechanism, machining chamber, electrolyte storage tank, pulse power supply unit, ultrasonic stack assembly and ultrasonic generator unit. A pulse power supply; in the voltage range of (6V – 30V), output current in the range of (0A - 33A), pulse on time range is available from 50μs to 2000μs in 12 steps and duty cycle from 8% to 96% in 12 steps are available. Research outcomes can be successfully implemented in the following fields:

a) Defense: The Defense Research and Development Organization (DRDO) and Bharat Heavy Electrical Limited (BHEL) which are engaged in turbine manufacturing can directly employ the above technique for drilling cooling holes in turbine blades made of super alloys where, high hardness and strength of material make them extremely difficult to machine.

b) Aviation Industry: Hindustan Aeronautics Limited (HAL) one of the oldest and largest aerospace industries engaged in manufacturing military aircraft can directly employ the above technique for machining engine components, rocket engine parts and jet engine rings.

c) Textile Industry: Spinnerets used in fiber manufacture require numerous tiny holes. Textile Industry can directly employ the above technique and produce tiny holes efficiently.

d) Automobile Industry: Firing chambers of internal combustion, gasoline engines.

e) Medical Applications: Producing pin holes in surgery components required during hip joint replacement and valve components that are made of titanium and cobalt based super alloy.

FIELD OF THE INVENTION

Our invention is related to:

- enhance the mass material removal rate (MRRg) and
- hole quality of hole produced on work-piece material made of titanium aluminide.
- Titanium aluminide is a potential light weight high strength temperature resistance alloy that can replace few of the high density materials used in aircraft, jet engines and automobiles components.
- design and develop a hybrid ultrasonic assisted pulse electrochemical machining setup for experimentation.
- Area of applied research: i)aviation, ii)automobile, iii)defense, iv) medical and v) textile industries.

BACKGROUND OF THE INVENTION

The schematic diagram of work procedure is displayed as Fig. 1. Pre-opening machining is done on 304 hardened steel plate first and foremost, which involves tube shaped twisting terminals for electrochemical penetrating interaction, the genuine image of cathode device for getting ready pre-opening.

To set up the pre-opening, it is important to choose sensible machining boundaries, which will fundamentally influence the opening breadth, opening roundness and opening divider surface unpleasantness of the pre-opening, and will influence the machining recompense during the opening developing cycle. Assuming the machining remittance of the pre-opening is too huge, the mechanical crushing will be excessively solid, bringing about low machining effectiveness and unfortunate surface nature of the inward mass of the opening.

Assuming the pre-opening machining stipend is too little, the mechanical crushing impact in UAECDG interaction will be debilitated, which won't altogether further develop the machining quality.

In this manner, the boundaries for setting up the pre-opening should be chosen sensibly, and the enhanced boundaries are displayed in Table 1. After pre-opening with certain machining precision and surface quality is machined, the prehole is broadened through ultrasonic-helped electrochemical drill-crushing.

PRIOR ART SEARCH

Electrochemical drill-crushing (ECDG) is a compound machining innovation, which consolidates electrochemical machining (ECM) with mechanical drill-crushing interaction. On this premise, another strategy for machining little openings which called ultrasonic-helped electrochemical drill-crushing (UAECDG) is proposed.

To begin with, the standard of UAECDG is dissected through examination of UAECDG process and electrochemical passivation conduct of materials. Second, the recreation of electrochemical drill-crushing interaction was examined to outline the impact of ball-end cathode on decreasing the opening shape and further developing the machining precision.

Thereafter, a few gatherings of analyses are led to examine the impact of electrical boundaries, ultrasonic plentifulness and matching degree among electrolysis and mechanical crushing on the machining nature of little openings.

At last, little openings with distance across of 1.1 ± 0.01 mm, surface unpleasantness of 0.31 lm and tighten of under 0.6 degree were machined by UAECDG, which uncovered UAECDG is a promising compound machining innovation to create little openings with top caliber and high productivity.

OBJECTIVES OF THE INVENTION

1. To develop hybrid ultrasonic assisted pulse electrochemical machining (USAPECM) setup.
2. To enhance the mass material removal rate (MRRg).
3. To produce holes on a work-piece material made of titanium aluminide material with dimensional accuracy.
4. To produce stress free (structural and thermal)cooling holes on titanium aluminide material.
5. To enhance the fatigue life of the machined components made of titanium aluminide material.

Research outcomes can be successfully implemented in the following fields:

- **Defense:** The Defense Research and Development Organization (DRDO) and Bharat Heavy Electrical Limited (BHEL) which are engaged in turbine manufacturing can directly employ the above technique for drilling cooling holes in turbine blades made of super alloys where, high hardness and strength of material make them extremely difficult to machine.
- **Aviation industry:** Hindustan Aeronautics Limited (HAL) one of the oldest and largest aerospace industries engaged in manufacturing military aircraft can directly employ the above technique for machining engine components, rocket engine parts and jet engine rings.
- **Textile Industry:** Spinnerets used in fiber manufacture require numerous tiny holes. Textile Industry can directly employ the above technique and produce tiny holes efficiently.
- **Automobile industry:** Firing chambers of internal combustion, gasoline engines.
- **Medical Applications:** Producing pin holes in surgery components required during hip joint replacement and valve components that are made of titanium and cobalt based super alloy.
- Composite, metal matrix composites.

Commercialization of inventions can lead to import substitutes.

SUMMARY OF THE INVENTION

Due to high mechanical strength of titanium aluminide conventional turning, drilling and milling is restricted as they produce various surface defects such as; micro cracks, voids, material distortion and heat affected zones. Damaged rim zone surface and poor surface quality are resulted due to mechanical and thermal stresses.

Electro discharge machining and laser beam machining produces heat affected zones and micro cracks on the work-piece. Ultrasonic machining causes tool wear and reduction in strength of work-piece.

The parts produced by electron beam machining required the secondary finishing operation. Secondary operation reduces the surface finish and restricts the use of these components in the aviation industry. Short circuiting, sparking and tool breakage in small inter electrode gap (IEG) restricts the use of direct current electrochemical machine (ECM) for deep hole drilling.

To circumvent frequent short-circuiting during ECM a new method, pulse electrochemical machining (PECM) was introduced in the year 2000. PECM improves the anodic dissolution efficiency, stable IEG and the better surface finish. ECM coupled with ultrasonic machining is used to disintegrate passive or non-conducting surface layers. Ultrasonic propagation effect in liquid media is called as ultrasonic cavitations which significantly influence electrochemical dissolution efficiency.

Various defects such as the formation of white layer, heat affected zone, change in microstructure, micro cracking, lesser surface integrity, accuracy and machining high aspect ratio holes.

The outcome of the invention are:

1. To minimize above machining defects and improve the dimensional accuracy a new USAPECM process is developed.
2. Enhancement in the mass material removal rate by 30%-40% as compared to direct current electrochemical machining (DCECM) and PECM.
3. Accuracy of drilled hole is enhanced by minimizing the taper. Diameter of the drilled hole at the entrance and exit found to be in close tolerance of ±5%.
4. Depth average radial overcut and uniform whole profile throughout is produced. Overcut is drastically reduced from 1.30 to 1.05.

BRIEF DESCRIPTION OF THE DIAGRAM

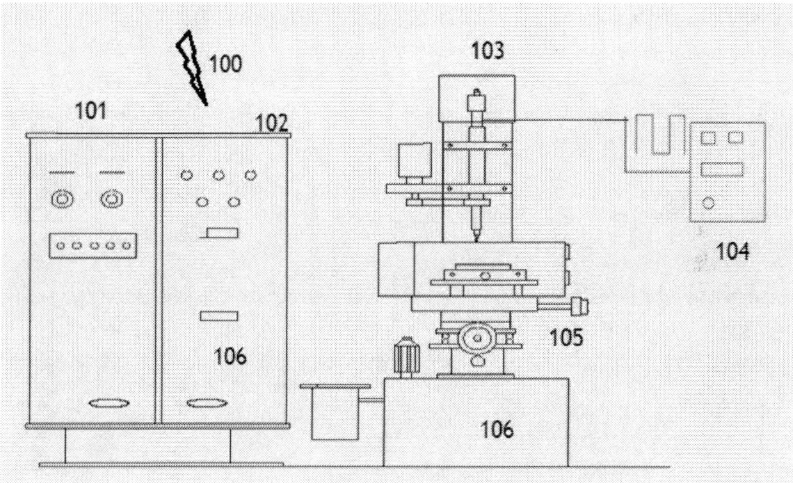

Fig.1: Enhancement in material removal rate and hole quality of titanium aluminide (light weight high strength temperature resistance) material using hybrid ultrasonic assisted pulse electrochemical machining process for industrial application Flow.

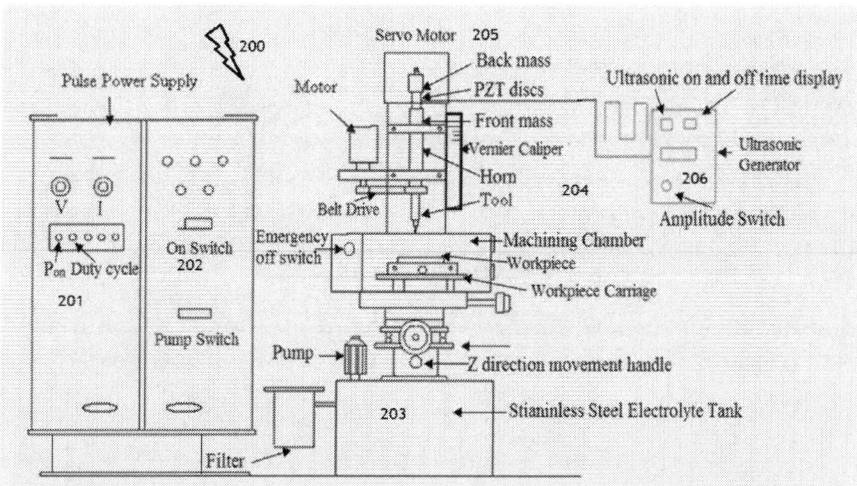

Fig. 2: Enhancement in material removal rate and hole quality of titanium aluminide (light weight high strength temperature resistance) material using hybrid ultrasonic assisted pulse electrochemical machining process for industrial application Block Diagram.

DESCRIPTION OF THE INVENTION

Pulse electrochemical machining process is hybridized with ultrasonic machining process and a new ultrasonic assisted pulse electrochemical machining setup is developed. The ultrasonic stack assembly capable of vibrating the electrochemical machining tool at 28-kHz frequency is successfully developed. Standard 230V, AC and 50Hz frequency electrical input is converted in to 28,000Hz, 300V electrical signal with the help of ultrasonic generator.

The main working elements of the ultrasonic assisted pulse electrochemical machine (USAPECM) are cathode tool, tool feed mechanism, machining chamber, electrolyte storage tank, pulse power supply unit, ultrasonic stack assembly and ultrasonic generator unit. A pulse power supply; in the voltage range of (6V – 30V), output current in the range of (0A - 33A), pulse on time range is available from 50 μs to 2000μs in 12 steps and duty cycle from 8% to 96% in 12 steps are available.

Electrolyte tank was fabricated in stainless steel (SS-304) with capacity of 175liters. Fresh electrolyte is charged in to the inter electrode gap (IEG) using chemical pump (0.5 HP, single stage, 230 V, single phase, 2.5A, 2900 rpm, min 11 and max 311/min). Machining of the titanium aluminide was performed in the machining chamber made up of SS 304 grade material. Work piece was held firmly using suitable work holding device.

Work piece was connected to the positive terminal and tool to negative terminal of power supply. The fresh electrolyte was charged at 1.2kg/cm2 between the IEG. Vertical feed to tool is obtained and maintained using Stepper motor and tool feed arrangement.

Detailed Description of the Invention

Titanium aluminide work piece and copper tool were connected to the positive and negative terminal of pulse power supply respectively.

Weight of the work-piece specimen was measured before and after completion of each experiment. Inter electrode gap (IEG) of 0.1mm between tool and work-piece was set using the slip gauges. Pulse power supply control parameters such as: pulse on time, pulse off time, duty cycle, voltage, gap current was set before each experiment. The power supply was put on before the electrolyte was pumped in the machining chamber to avoid formation of the passivation layer. The NaNO3 electrolyte from a stainless-steel tank was pumped in to IEG through a filter.

Electrolyte conductivity was measured using a digital electrolyte conductivity meter with cell constant of 0.98. The ultrasonic generator was tuned to the power and the required parameters such as ultrasonic on time, ultrasonic off time, duty cycle and amplitude of ultrasonic vibration were set. Constant voltage transformer was used to supply constant voltage to the ultrasonic generator to avoid failure of the transducer and other components of stack assembly due to fluctuation in voltage.

In all the experiments, holes (3mm diameter x 3mm deep) were drilled. The time required for machining was noted using a stopwatch to evaluate the mass material removal rate (MRRg). To obtain a through hole, dummy plate of 0.5 mm thickness was kept below the work piece. Once a hole is drilled the machine was stopped and the work piece is removed from the machining chamber. The work-piece was washed using distilled water, dried and weighed using a digital weighing machine with a least count of 0.01gm.

Measurements were carried out on the machined hole after each experiment.

- To measure the mass material removal rate, a weight loss method was used. The time required for machining was noted using precision stopwatch (Baker India). Weight of the work-piece before and after each experiment was noted under the dry condition on weighing machine (SHIMADZU), with least count (0.01gm).
- Diameters were measured at the five different positions along the depth of the machined hole (at 0.5mm interval from top face of work-piece) using a coordinate measuring machine (SPECTRA) with 1mm ruby ball diameter stylus. The machine was calibrated to the known specimen, before the actual measurements were performed. The machine was computerized numerical control operated and capable of obtaining the results with ± (2.5+ L/350) microns accuracy.

OUTCOME

Pulse electrochemical machining process is hybridized with ultrasonic machining process and a new ultrasonic assisted pulse electrochemical machining setup is developed. The ultrasonic stack assembly capable of vibrating the electrochemical machining tool at 28-kHz frequency is successfully developed. Standard 230V, AC and 50Hz frequency electrical input is converted into 28,000Hz, 300V electrical signal with the help of an ultrasonic generator. The main working elements of the ultrasonic assisted pulse electrochemical machining (USAPECM) setup are cathode tool, tool feed mechanism, machining chamber, electrolyte storage tank, pulse power supply unit, ultrasonic stack assembly and ultrasonic generator unit. A pulse power supply; in the voltage range of (6V – 30V), output current in the range of (0A - 33A), pulse on time range is available from 50μs to 2000μs in 12 steps and duty cycle from 8% to 96% in 12 steps are available. Research outcomes can be successfully implemented in the following fields: Pulse electrochemical machining process is hybridized with ultrasonic machining process and a new ultrasonic assisted pulse electrochemical machine setup is indigenously developed.

- According to claim1# the invention is to provide a pulse electrochemical machining process is hybridized with ultrasonic machining process and a new machine ultrasonic assisted pulse electrochemical machine is developed and also the main working elements of the ultrasonic assisted pulse electrochemical machine (USAPECM) are cathode tool, tool feed mechanism, machining chamber, electrolyte storage tank, pulse power supply unit, ultrasonic stack assembly and ultrasonic generator unit.
- According to claim 1,2# a pulse power supply; in the voltage range of (6V – 30V), output current in the range of (0A - 33A), pulse on time range is available from 50μs to 2000μs in 12 steps and duty cycle from 8% to 96% in 12 steps are available and also The ultrasonic stack assembly capable of vibrating the electrochemical machining tool electrode at 28 kHz frequency is successfully developed. Standard 230V, AC and 50Hz frequency electrical input is converted into 28,000Hz, 300V electrical signal with the help of an ultrasonic generator.
- According to claim 1,2,3# research outcome can be directly implemented in various industrial applications such as: Defense: The Defense Research and Development Organization (DRDO) and Bharat Heavy Electrical Limited (BHEL) which are engaged in turbine manufacturing can directly employ the above technique for drilling cooling holes in turbine blades made of super alloys where, high hardness and strength of material make them extremely difficult to machine. Automobile industry: Firing chambers of internal combustion, gasoline engines. Textile Industry: Spinnerets used in fiber manufacture require numerous tiny holes. Textile Industry can directly employ the above technique and produce tiny holes efficiently.

- According to claim 1,2,3,4# Medical Applications: Producing pin holes in surgery components required during hip joint replacement and valve components that are made of titanium and cobalt based super alloy. Aviation industry: Hindustan Aeronautics Limited (HAL) one of the oldest and largest aerospace industries engaged in manufacturing military aircraft can directly employ the above technique for machining engine components, rocket engine parts and jet engine rings. Commercialization of invention can lead to import substitute.
- According to claim 1, 2,# the invention is capable of vibrating the tool electrode made of copper material at 28kHz frequency in lateral mode of vibration and also the invention is capable of machining the work piece under stationary position (i.e. non rotating) mode and also The invention is capable of vibrating the tool electrode at four amplitude 25µm, 50 µm, 75 µm and 100 µm.
- According to claim 1,2,3,4,5,# the machining was performed by clamping the work-piece in a fix position and also The designed and developed hybrid ultrasonic assisted pulse electrochemical machine is capable of producing the high aspect ratio hole up to 1:20.
- According to claim 1,2,3# the mass material removal rate of USAPECM machine found to be enhanced by 30%-40% as compared to direct current electrochemical machining (DCECM) and pulse electrochemical machining (PECM) and also According to claim 1,2,3,4,5,6,7 # the hole taper found drastically reduced in USAPECM as compared to DCECM and PECM. Diameter of the drilled hole by USAPECM process found to be in close tolerance of ±5%. at the entrance and exit.
- According to claim 1,2,3,4,5,6,7,8# USAPECM machining process drastically reduces the overcut from 1.30 to 1.05.

CITATIONS

- 202022100309, Granted, Valectus Private Limited, Method for Storage and Processing of Geospatial Data for Efficient Rendering and Visualization on a Server, : 18-B, Gurudev Apartments, 1st Floor, Chembur Naka, R C Marg, Chembur, Mumbai, Mumbai City, Maharashtra, India, 400074.
- 202021106919, Granted, Dr. Daphne Pillai, Dr. Priam Pillai , Franav Pillai, Deepika Sharma, Dr. Kavita Kathare, Skill Credit System, MAHATMA EDUCATION SOCIETY'S, Pillai College of Arts Commerce & Science (Autonomous).
- 202221001647, MAHATMA EDUCATION SOCIETY'S, Pillai College of Arts Commerce & Science, White Board "Dry and Wet" Flip Duster Eraser" Dr. Gajanan Wader, Principal, Dr. Shabab Ahmed Rizvi, Assistant Professor.
- 202221001646, Dr. Aarti Sukheja, Dr. Kavita Kathare, Intelligent Air Conditioning System Notify The part that needs repair or replacement, MAHATMA EDUCATION SOCIETY'S, Pillai College of Arts, Commerce & Science (Autonomous).
- 202221001645, Dr. G R Patil , Dr. K D Singh ,Prof. Raju Narwade, Prof. Karthik Nagarajan., Prof. Manisha Jamgade Dual triangular slope bottom tuned liquid damper for improving efficiency of flat bottom tuned liquid damper. MAHATMA EDUCATION SOCIETY'S, PILLAI HOC COLLEGE OF ENGINEERING &TECHNOLOGY RASAYANI.

Chapter - 10

Street Vendors Study Using Machine Learning

Dr. Jaya Cherian

Asst. Prof, MES's Pillai College of Education and Research, Chembur

Dr. Reni Francis

Principal, MES's Pillai College of Education and Research, Chembur

Dr. Vineet Tirth

(King Khalid University Abha Saudi Arabia)

Mechanical Engineering Department, College of Engineering, King Khalid University, Abha 61421, Asir, Kingdom of Saudi Arabia

ABSTRACT

Our Invention "Street Vendors Study using Machine Learning Is a The experiences of street vendors in Mumbai, capital Money Market of India's, are examined in this research. The purpose of the study was to learn more about street vendor livelihoods in the city, with a special focus on barriers to livelihood security. The study investigates a situation in which vendors face ongoing issues such as forceful removals, exploitation, and extortion. The invention's second goal is to figure out how vendors may band together to address these difficulties together. Although other kinds of labour organisations exist, the study contends that they are often reactive and limited to regions with a well-established vending community. This article compiles recent research and evidence on urban policy and local government practices in relation to street vendors, one of the most conspicuous informal economy jobs. This study is based on qualitative techniques applying case studies, observations methods and interview used for investigate the causes for evictions and relocations, this research uses participatory methods and observation methods to investigate the more "everyday" issues that street vendors face. This study will help to understand economic condition of street vendors and also help government to make a policy and procedures regarding street vendors. This research investigates the regulatory settings in which local government officials have created these informal practices, as well as the extent to which street sellers have access to public space.

BACKGROUND OF THE INVENTION

Customarily, road dealers are remembered to work outside of state administrative frameworks. Ongoing examination, then again, plays featured the state's part in laying out merchants' casual status and recorded neighborhood government exercises that exploit a mistaking lawful climate for sellers. Low-level badgering, products seizures, and inconsistent removals are instances of these strategies.

Road merchants and market venders are a significant element of worldwide metropolitan economies, giving simple admittance to a different scope of reasonable labor and products in open areas.

They offer everything from new produce to arranged dishes, development materials to attire and specialties, and customer hardware to auto fixes and hair styles. Not in the least do road dealers make occupations for themselves, yet additionally for doormen, safety officers, transportation administrators, and capacity providers.

Installments for licenses and allows, expenses and fines, and a few sorts of charges are for the most part ways for urban areas to bring in money. In urban communities, road venders make up a huge and apparent labor force, yet it's difficult to evaluate their numbers unequivocally.

PRIOR ART SEARCH

Advancement is a multi-layered peculiarity. A portion of its significant aspects include: the degree of financial development, level of training, level of wellbeing administrations, level of modernization, status of ladies, level of nourishment, nature of lodging, appropriation of labor and products, and admittance to correspondence.

In India, the advancement of financial improvement among significant states isn't uniform. This study inspects the current inconstancy of between state advancement and accordingly distinguishing the pointers liable for the variety being developed.

Rather than concentrating on the inconstancy of a specific variable across states, a composite record in light of a few markers has been created utilizing head part investigation and states are organized by the files determined utilizing four comprehensively acknowledged parts:

1. Financial creation and monetary condition or as such degree of financial turn of events;
2. Normal least requirements;
3. Wellbeing and wellbeing related administrations
4. Correspondence.

The discoveries of the investigation support the overall insight about the states. The states in India are set apart with wide difference in financial turn of events. The variables, which are found out to be more significant for the general advancement process, connect with fundamental requirements like instruction, accessibility of food, least buying power and offices like safe drinking water, medical services framework, and so forth

It is likewise observed that enlistment proportion can't be raised except if least necessities of the average citizens are fulfillcd. Along these lines, genuine advancement requires government activity to work on rudimentary training, safe drinking water offices and medical care, and to eliminate obstructions against social minorities, particularly ladies.

The job of social improvement like proficiency (and especially of female education) in advancing essential abilities arises as the essential to in general turn of events. These outcomes obviously accentuate the job of well-working public activities in working on the general day to day environments of individuals.

Albeit monetary development in the feeling of growing gross public item and other related factors is one of the most major contribution to the general advancement process, the essential goal of improvement should zero in on the extension of human capacities which has been dismissed for long in India.

OBJECTIVES OF THE INVENTION

- The objective of the invention is to provide a have a better understanding of their business activity's related to financial accessibility.
- The other objective of the invention is to provide a investigate how street vendors achieve social security.
- The other objective of the invention is to provide a investigate the occupational wellbeing of street vendors, taking into

account their income level, working hours, workplace safety, bribe payment, workplace dignity, and the role of government in improving their 'working environment'.

- The other objective of the invention is to provide a "Socio-economic realities of the outliers of India's Urban Landscape: Mixed research study on street vendors." Is a The experiences of street vendors in Mumbai, capital Money Market of India's, are examined in this research.
- The other objective of the invention is to provide a purpose of the study was to learn more about street vendor livelihoods in the city, with a special focus on barriers to livelihood security. The study investigates a situation in which vendors face ongoing issues such as forceful removals, exploitation, and extortion.
- The other objective of the invention is to provide a invention's second goal is to figure out how vendors may band together to address these difficulties together. Although other kinds of labour organisations exist, the study contends that they are often reactive and limited to regions with a well-established vending community.
- The other objective of the invention is to provide a article compiles recent research and evidence on urban policy and local government practices in relation to street vendors, one of the most conspicuous informal economy jobs.
- The other objective of the invention is to provide a study is based on qualitative techniques applying case studies, observations methods and interview used for investigate the causes for evictions and relocations, this research uses participatory methods and observation methods to investigate the more "everyday" issues that street vendors face.
- The other objective of the invention is to provide a study will help to understand economic condition of street vendors and also help government to make a policy and procedures regarding street vendors.
- The other objective of the invention is to provide a research investigates the regulatory settings in which local government officials have created these informal practices, as well as the extent to which street sellers have access to public space

SUMMARY OF THE INVENTION

After the Second World War, the investigation of monetary advancement of the monetarily in reverse nations became quickly with the issue of financial requirements and hardship of the average folks as the focal significance (Hall, 1983). Concentrates in this regard and thusly in the space of human turn of events and actual personal satisfaction, particularly with regards to India, are voluminous (Sen, 1981; Ram, 1982; Karkal and Irudaya Rajan, 1991; Nissan, 1993).

A few creators utilized the system given by UNDP to build the composite file of improvement. Independent of the contentions and discussion in regards to the techniques to seek after the arrangement, there stays the topic of how to quantify the advancement of improvement.

As the significant focal point of this invention is to study the financial differential and related conduct in a formative viewpoint, the full scale level nation attributes and individual state level factors influencing personal satisfaction, straightforwardly or by implication, through formative interaction must be recognized. "The Indian subcontinent, with its enormous size, wide construction and eco-social inconsistencies is better perceived and better deciphered when learned at the territorial level.

Examination of information at disaggregated structure limits the inconstancy and empowers better distinguishing proof of exceptional qualities" (Dutta Roy Choudhury, 1995). Likewise, the nation is isolated into five distinct locales of 16 significant states, which represent over 95% of the all-out populace.

The Northern area is involved four states: Punjab, Haryana, Himachal Pradesh, and Rajasthan; western locale is included two states: Maharashtra and Gujarat; eastern district is contained four states: West Bengal, Assam, Orissa and Bihar; and the focal locale is involved two states: Uttar Pradesh and Madhya Pradesh; lastly southern area is involved four states: Andhra Pradesh, Kerala, Tamil Nadu, and Karnataka.

To get a composite record from a bunch of factors, a wide assortment of multivariate factual strategies are accessible. As a matter of fact, the decision of the most suitable strategy relies on the sort of the issue, the idea of the information and the target of the examination. In sociology, factors are, by and large, related and the scientists are not in a situation to concentrate on the financial elements with a bunch of free factors.

One requirements to search for an elective aspect decrease strategy which will empower them to sum up the entire arrangement of data into a sensible structure absent a lot of loss of the data content of the first information. The topic of the multivariate examination is rearrangements and "to sum up an enormous assemblage of information through somewhat couple of boundaries" (Chatfield and Collins, 1980).

However the composite file can be developed utilizing basic strategies like positioning and ordering techniques, these methods have numerous downsides which have been reprimanded by numerous analysts like Dander Committee (1984), Kunduz and Raza (1982), and Sarker (1995). Fundamental downsides are assertion and designation of equivalent loads.

Kendall (1939) fostered a composite file equation to beat these issues utilizing between subordinate factors relating to horticultural efficiency, which is officially comparable to what Hoteling (1933) called as a "first head part". In head part examination (PCA), a bunch of unique factors is changed to another arrangement of uncorrelated factors called head parts.

These new factors are straight elements of the first factors and determined in diminishing request of significance. The goal is to figure out a couple of parts, which represent the vast majority of the variety in the first informational collection. For subtleties of approach, allude to Anderson (1985), and Chatfield and Collins (1980).

BRIEF DESCRIPTION OF THE DIAGRAM

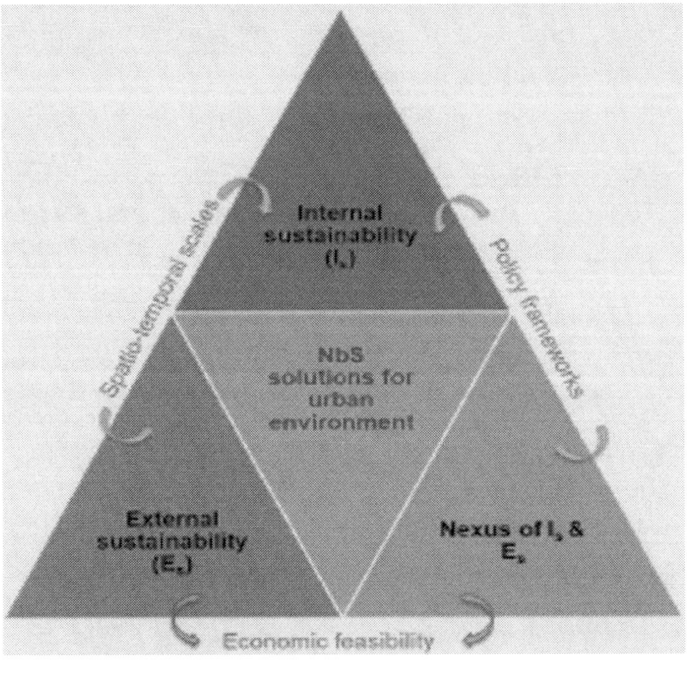

Fig.1: Socio-economic realities of the outliers of India's Urban Landscape: Mixed research study on street vendors flow

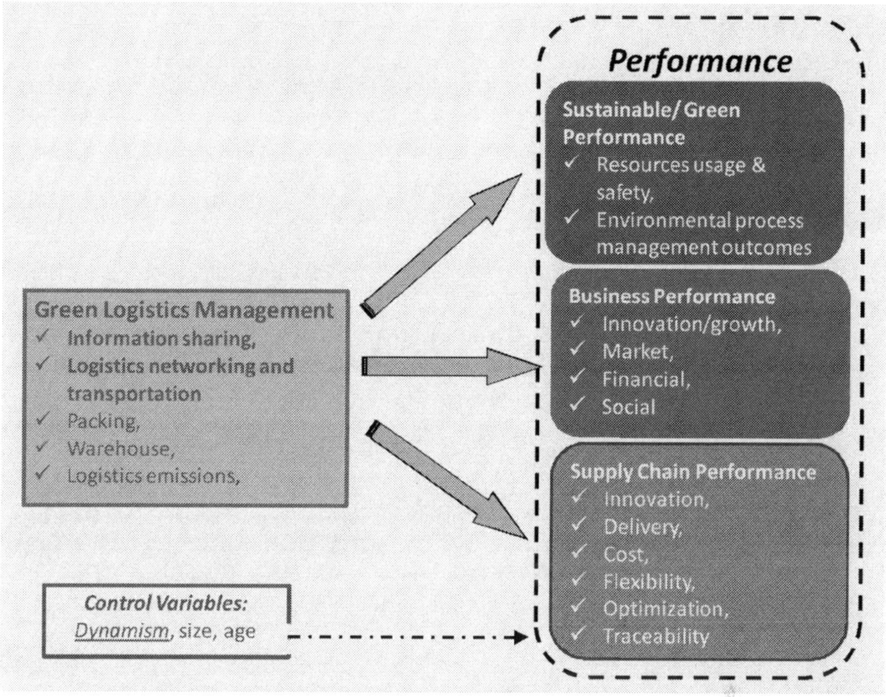

Fig. 2: Socio-economic realities of the outliers of India's Urban Landscape: Mixed research study on street vendors block diagram

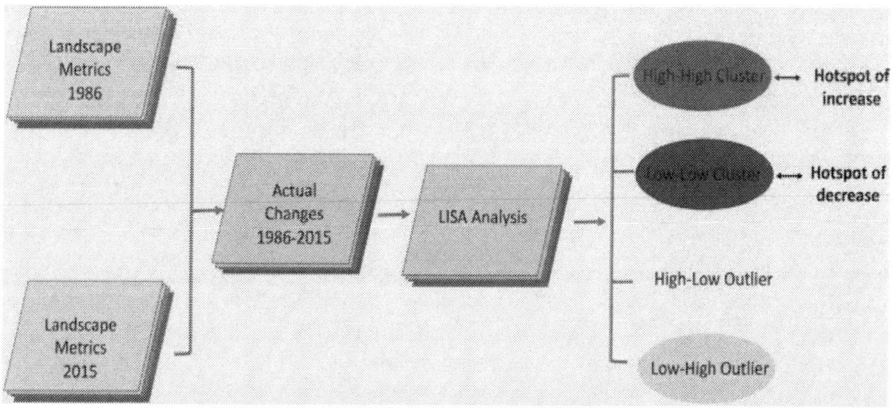

Fig.3: Socio-economic realities of the outliers of India's Urban Landscape: Mixed research study on street vendors process

DESCRIPTION OF THE INVENTION

Those figures have been utilized here. A few arranging procedures are being carried out to address the difficulties to kill destitution. It is seen that the chief need in destruction of neediness is to meet the normal essential requirements of individuals. "Neediness must be related to lack in complete degree of living.

What's more, absolute degree of living incorporates energy prerequisites as well as adjusted diet required for wellbeing, and different parts of fundamental necessities fundamental for human life at an average level" (Rao, 1977). The level of individuals underneath the destitution line of the state shows their degree of result, pay and appropriation of vital labor and products independent of the discussion of deciding the neediness line.

However there is no broad understanding about a "ordinary" calorie admission, we have taken the state level information on level of individuals above neediness line from CMIE (Center for Monitoring Indian Economy) distribution on friendly areas.

First head part scores of individual states have been organized in dropping request and introduced in. Out of 16 significant states, 7 states were found to perform over the normal and the scores in the primary head part changed between - 2.26 to 2.99. Punjab and Himachal Pradesh, with their particular scores of 2.99 and 2.18, remained at the top.

Interestingly, Assam was at the lower part of the ordered progression. Truth be told there was essentially little distinction among the last three states, Assam, Orissa and Bihar, in this gathering. West Bengal was introduced just beneath the normal norm. The vast majority of the southern states were greatly improved contrasted with eastern and focal territorial states in India.

OUTCOME

Our Invention "Socio-economic realities of the outliers of India's Urban Landscape: Mixed research study on street vendors." Is a The experiences of street vendors in Mumbai, capital Money Market of India's, are examined in this research. The purpose of the study was to learn more about street vendor livelihoods in the city, with a special focus on barriers to livelihood security. The study investigates a situation in which vendors face ongoing issues such as forceful removals, exploitation, and extortion. The invention's second goal is to figure out how vendors may band together to address these difficulties together. Although other kinds of labour organisations exist, the study contends that they are often reactive and limited to regions with a well-established vending community. This article compiles recent research and evidence on urban policy and local government practices in relation to street vendors, one of the most conspicuous informal economy jobs. This study is based on qualitative techniques applying case studies, observations methods and interview used for investigate the causes for evictions and relocations, this research uses participatory methods and observation methods to investigate the more "everyday" issues that street vendors face. This study will help to understand economic condition of street vendors and also help government to make a policy and procedures regarding street vendors. This research investigates the regulatory settings in which local government officials have created these informal practices, as well as the extent to which street sellers have access to public space.

- According to claim1# the invention is to a "Socio-economic realities of the outliers of India's Urban Landscape: Mixed research study on street vendors." Is a The experiences of street vendors in Mumbai, capital Money Market of India's, are examined in this research.
- According to claim1,2# the invention is to a purpose of the study was to learn more about street vendor livelihoods in the city, with a special focus on barriers to livelihood security. The study investigates a situation in which vendors face ongoing issues such as forceful removals, exploitation, and extortion.
- According to claim1, 2,4# the invention is to a invention's second goal is to figure out how vendors may band together to address these difficulties together. Although other kinds of labour organisations exist, the study contends that they are often reactive and limited to regions with a well-established vending community.
- According to claim1,2,3,4# the invention is to a article compiles recent research and evidence on urban policy and local government practices in relation to street vendors, one of the most conspicuous informal economy jobs.
- According to claim1,2,4,5# the invention is to a study is based on qualitative techniques applying case studies, observations methods and interview used for investigate the causes for evictions and relocations, this research uses participatory methods and observation methods to investigate the more "everyday" issues that street vendors face.

- According to claim1,2,3,4,5# the invention is to a study will help to understand economic condition of street vendors and also help government to make a policy and procedures regarding street vendors.
- According to claim1,2,4,5,6# the invention is to a research investigates the regulatory settings in which local government officials have created these informal practices, as well as the extent to which street sellers have access to public space

CITATIONS

- 202221006893, Esakkiammal Sudalai Pandi Nadar, MUMBAI BEST BUS TRACKER, MES, Mahatma Junior College of Teacher Education, Opposite, Fire Brigade, Sion Trombay Road, Chembur Naka, Mumbai 400071, India.
- 202221008799, Dr. Ashish Tripathi, Ms. Shilpa Tripathi, Ms. Agrima Tripathi, Without any Software I Have Built an Automated ATTENDANCE System, MAHATMA EDUCATION SOCIETY'S, PILLAI COLLEGE OF ENGINEERING Management.
- 202221010401, Dr. Mansi Subhedar ,Dr. B. K. Sarkar, LED Dimming Driver Circuit Using PWM MAHATMA EDUCATION SOCIETY'S, PILLAI HOC COLLEGE OF ENGINEERING & TECHNOLOGY RASAYANI, TALUKA PANVEL.
- 202221010400, Prof. (Dr.) Shilpa Pankaj Kewate, 3 Bit Error-Correcting Modified Linear Block Code MAHATMA EDUCATION SOCIETY'S, PILLAI HOC COLLEGE OF ENGINEERING & TECHNOLOGY RASAYANI, TALUKA PANVEL.

Chapter - 11

Tilt-Resistant Rover Using Machine Learning

Prof (Dr.) B.K. Sarkar

MES's, Pillai HOC College of Engineering & Technology, Rasayani Taluka Panvel, Dist, Navi Mumbai, Maharashtra 410207, India.

Anish Kumar

H No 1627, Sec- 5, Urban Estate, Kurushetra, Haryana 136118, India.

Prof. (Dr.) S. B. Chordiya

(DIRECTOR-SIMMC-CAMPUS) Suryadatta Institute of Management & Mass Communication (SIMMC), Sr. No: 342, Bavdhan, Pune-411021, Mh, India.

ABSTRACT

Our Invention "Tilt-Resistant Rover using Machine Learning " is a mission of ISRO in 2019 when the Vikram lander couldn't land as expected because of programming blunder I found out about making the plan so that can keep it protected and stable by utilizing round rings that don't permit it to shift or even to get topsy turvy. As I expressed above, I tried it thoroughly for itself and was effective and furthermore Mobile robots are of exorbitant interest for automated planetary investigation. The extremely fruitful Pathfinder mission to Mars has stunningly shown the capability of portable stages for planetary investigation. The European Space Agency (ESA) likewise began to foster ideas for miniature wanderers for Mars missions. Inside an interdisciplinary gathering of organizations had practical experience in space applications and examination labs new plans of miniature wanderers have been explored. Two ideas, a basic and hearty one and an inventive one, have been chosen and useful breadboard models of them are at present fabricated. After a conversation of the main points of interest for hearty motion the current paper will zero in on the plan and control of the more creative arrangement. It comprises of 4 autonomously determined wheels organized in two triangles. It in this manner permits for effective moving on level surfaces as well as to step on deterrents. Furthermore the focal point of mass and the instrumentation carrousel is customizable, permitting to ideally adjust the miniature meanderer in pretty much every circumstance. Indeed, even subsequent to flipping over the robot can continuously get back on its wheels.

FIELD OF THE INVENTION

Field involves space sciences, science and technology, outer-space communication, electronics, mechatronics, Robotics.

BACKGROUND OF THE INVENTION

For the future planetary missions to Mars and Moon the European Space Agency (ESA) is looking for miniature meanderers to investigate a region with a sweep of the request for 10 m around the lander. Itemized foundation data can be found in.

This miniature wanderer ought to have the option to move to objects, to make estimations with its ready instruments and to send the data back to the lander. To concoct new creative arrangements a group of organizations and exploration labs was chosen. The objective of the venture is to foster new ideas, to choose the two most encouraging ones.

- basic and strong one and a more imaginative one)
- to fabricate models for check of their exhibitions.

OBJECTIVES OF THE INVENTION

- The objective of the invention is to provide a 1 It will not tilt or get upside down in any case with help of circular rings
- The other objective of the invention is to provide a 2 As claimed in claim1 I concentrated the mass at a single point
- The other objective of the invention is to provide a 3 To make the 2nd claim possible I used the inverted 'v' shape on sides to attach wheels.

- The other objective of the invention is to provide a 4 To make claim 1 possible in storms I used a storm detection system
- The other objective of the invention is to provide a 5 As claimed in claim-4 to protect vehicle from storm, there will be two autonomous arms installed on it

SUMMARY OF THE INVENTION

My principle center is to make the wanderer so that the meanderer won't turn over regardless until some other power is applied by a person. In the event that it slants towards the sides, 2 arms will work to return it to the way so that the fold will be pushed in outward heading returning the wanderer to its unique position.

The tempest identification framework works so that, which will be associated with an engine connected with a fan/turbine which will create some voltage assuming the tempest strikes. In response an arm will fall on a superficial level to give grasp to the vehicle.

Requirements

Inside the structure of the current advancement ESA characterized the accompanying primary necessities and limitations:

- Stowed aspects (cm): 30 x 20 x 20
- Net mass of the robot framework: max. 2 kg
- Mass of the logical payload: 2 kg
- Electrical power given via lander: max. 2 Watt normal, at the most. 3 Watt top
- Maximum impediment tallness to survive: 10 cm
- Possibility to situate the on board sensors: exactness around 1 mm
- Maximum speed: 5 m/h the meanderer should conquer hindrances of 0.1 m stature and openings of 0.1 m width. It must have the option to ascend slants of 15° and down 20°.

Straightforward moves like turning and going in reverse is required as well. These determinations connected with the territory can be utilized as a benchmark. To guarantee usefulness in genuine Martian territory, expanded tests in different conditions will be made. To guarantee magnificent execution the accompanying significant issues have been recognized and determined:

- No gamble of sinkage and stalling out in the sand
- Flip-over dependability, strength against wind
- Recovery after flip-over and subsequent to having been covered by a dust storm

Turning should be possible by applying two distinct rates to the wheels. Grinding on the wheels is significantly diminished either by turning the two wheel-outlines so just a single wheel is in ground contact at the mark of revolution or by changing the focal point of gravity to the fitting spot near the front wheels.

With the idea right B each of the 4 wheels are kept up with on the ground, yet the payload is uprooted so the focal point of gravity is basically on the two front wheels. The different turn speeds on the two front wheels arrange the direction and the lingering low powers on the two others wheels will lessen the erosion on ground to an extremely low level.

Another vital angle concerns the capacity to recuperate subsequent to flipping over. By moving the fundamental mass out, the focal point of gravity can be put external the base framed by the three wheels. This permits to bring the meanderer back on its wheels.

CONTROL OF THE TILT-RESISTANT ROVER

Control is thought to be circulated over the meanderer, the lander and ground (Earth). Different models and ideas, some of them more progressive and some of them more way of behaving based have been proposed for space robots. Our methodology fits inside the brought together control design for planetary wanderers which has been proposed in.

The level of independence of a robot depends straightforwardly of the intricacy of its neighborhood control. Absolutely independent portable frameworks are exceptionally difficult to fabricate on account of energy and route issues. For our situation, a semi-independent robot is more adjusted to satisfy the mission's necessities. A significant part of the entire framework is the low transmission rate from Mars to Earth.

A tele operator ought to act just for way arranging and if there should be an occurrence of perils. Consistently, all choices are made by the lander and additionally the wanderers. Neighborhood insight should along these lines surpass the reflexive ways of behaving and give critical thinking to more perplexing circumstances.

BRIEF DESCRIPTION OF THE DIAGRAM

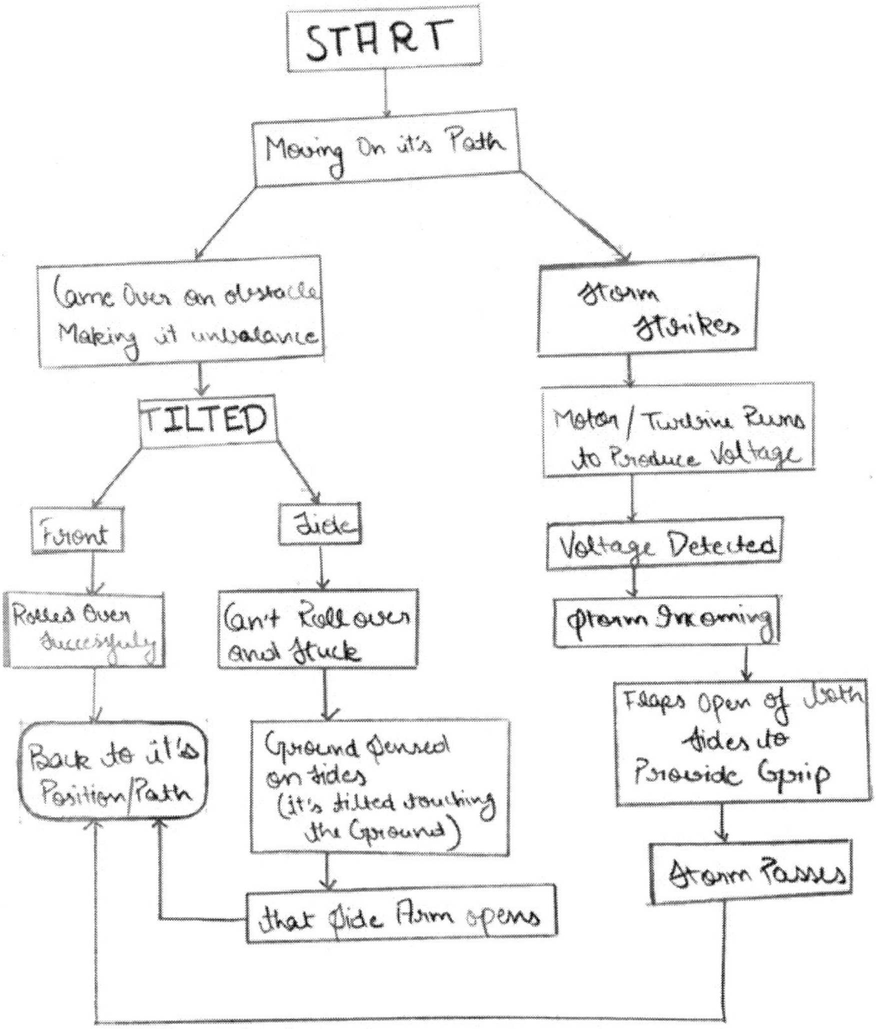

Fig. 1: Tilt-resistant rover flow chart

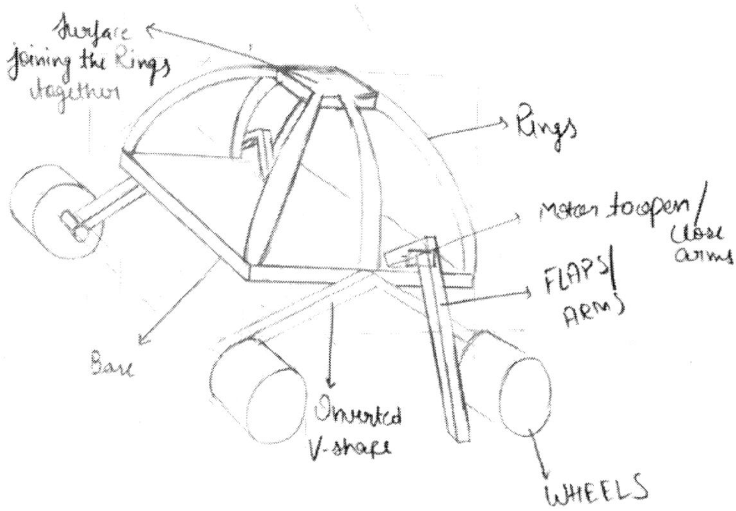

Fig. 2: Tilt-resistant rover, block diagram

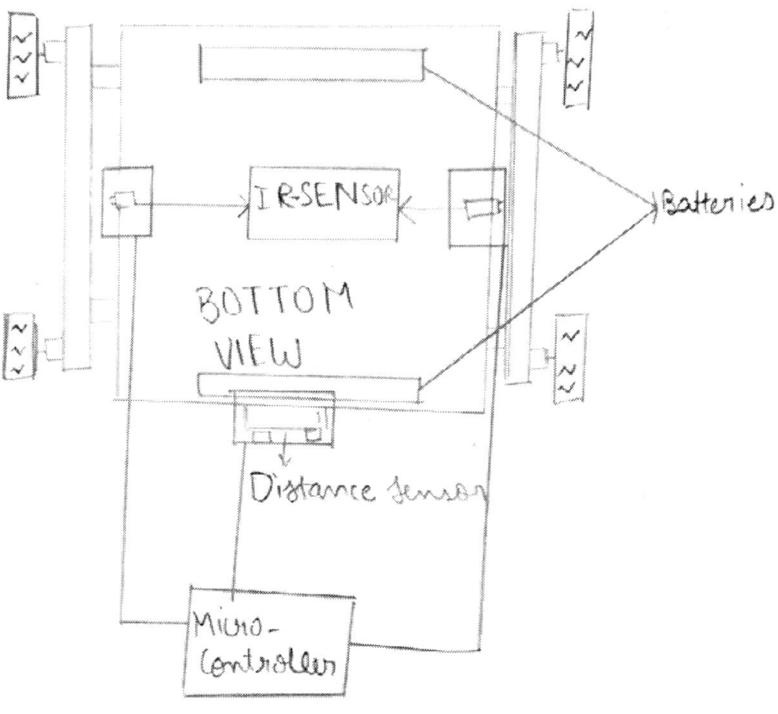

Fig. 3: Tilt-resistant rover, Block Diagram

DESCRIPTION OF THE INVENTION

Four primary standards - rolling, strolling, slithering and bouncing - have been recognized for full or halfway strong state contact. In any case, extra movement standards without strong state contact could be of interest in extraordinary climate. The greater part of the versatile robot for planetary investigation will move the vast majority of their experience on almost level surfaces, where moving movement has its most elevated proficiency and execution.

Be that as it may, some crude climbing capacities are expected as a rule. Along these lines mixture draws near, where for instance moving movement is joined with venturing, are of exorbitant interest. Gives an outline of qualities of the different motion ideas. The scoring addresses our closely held individual belief and is obviously not impartial.

As should be visible, the moving headway has just little detriments, primarily concerning the conquering of articles. With the proposed mixture approach this flimsy part is settled, yet the strength is brought down. The 'Venturing Triple Wheels' which is additionally viewed as in the table was chosen as the most encouraging contender for the creative arrangement.

Mechanical Design

Our locomotion concept is a hybrid approach called stepping triple wheels (figure 1) in view of both, strolling and moving abilities like AIMARS. Two autonomously determined sets of three wheels are upheld by two casings. The casings can turn freely around the fundamental body (payload outline) and permit the wanderer to effectively lift one wheel to step on objects.

The mechanical plan of the meanderer can be isolated in two distinct parts, the wheel outline and the payload outline, which can be planned autonomously. The initial segment respects the haggles outline which support them. The subsequent part concerns the logical payload and its help. Concerning the wheel approaches at least four levels of opportunity (DOF) are required. Two of them are need to permit the vehicle to move in the plane.

To give a crude strolling capacity two extra levels of opportunity are added. There are a great deal of conceivable outcomes to execute such a system. In the introduced plan 8 engines are utilized so that each haggle can be driven autonomously. The benefits are high overt repetitiveness and a less difficult plan. Right show the three haggles four engines of the wheel outline.

The payload edge of the meanderer contains every one of the logical instruments important to achieve the mission. With 2 kg the payload is the main piece of the complete wanderer mass. By moving the payload outline generally to the wheel approaches the focal point of gravity of the wanderer can be changed. This adds new abilities for headway, for example to jump on deterrents or to recuperate subsequent to flipping over.

Electronic Design

To control the various engines and sensors a suitable electronic should be incorporated in the meanderer. Turning contacts (slip rings) are utilized to interface the singular circuits on the casings. Each turning contact has two leads for power supply and two for a sequential connection point.

The itemized engineering of the electronic sub-framework is depicted in the plan. The hardware of the robot framework is made out of four isolated modules which imparts each other by means of a sequential connection. Three modules are accountable for the engines and sensors control. The job of the last module is to arrange the others, to take more elevated level choices and to speak with the lander.

WE CLAIMS

Our Invention “Design and development of tilt-resistant rover using autonomous arms “ is a mission of ISRO in 2019 when the Vikram lander couldn’t land as expected because of programming blunder I found out about making the plan so that can keep it protected and stable by utilizing round rings that don’t permit it to shift or even to get topsy turvy. As I expressed above, I tried it thoroughly for itself and was effective and furthermore Mobile robots are of exorbitant interest for automated planetary investigation. The extremely fruitful Pathfinder mission to Mars has stunningly shown the capability of portable stages for planetary investigation. The European Space Agency (ESA) likewise began to foster ideas for miniature wanderers for Mars missions. Inside an interdisciplinary gathering of organizations had practical experience in space applications and examination labs new plans of miniature wanderers have been explored. Two ideas, a basic and hearty one and an inventive one, have been chosen and useful breadboard models of them are at present fabricated. After a conversation of the main points of interest for hearty motion the current paper will zero in on the plan and control of the more creative arrangement. It comprises of 4 autonomously determined wheels organized in two triangles. It in this manner permits for effective moving on level surfaces as well as to step on deterrents. Furthermore the focal point of mass and the instrumentation carrousel is customizable, permitting to ideally adjust the miniature meanderer in pretty much every circumstance. Indeed, even subsequent to flipping over the robot can continuously get back on its wheels.

- According to claim1# the invention is to a 1 It will not tilt or get upside down in any case with help of circular rings
- According to claim1,2# the invention is to a 2 As claimed in claim1 I concentrated the mass at a single point
- According to claim1,2,3# the invention is to a 3 To make the 2nd claim possible I used the inverted 'v' shape on sides to attach wheels.
- According to claim1,2,3,4# the invention is to a 4 To make claim 1 possible in storms I used a storm detection system.
- According to claim1,2,4# the invention is to a 5 As claimed in claim-4 to protect vehicle from storm, there will be two autonomous arms installed on it.

CITATIONS

- 202221001660, Dr. Biplab Kumar Sarkar, Investigating the Cases of Novel Coronavirus Disease (Covid-19) Using Intelligent Dynamic Statistical Techniques, MAHATMA EDUCATION SOCIETY'S, PILLAI HOC COLLEGE OF ENGINEERING & TECHNOLOGY RASAYANI, TALUKA PANVEL.
- Dr. Biplab Kumar Sarkar, 202241000869, MAHATMA EDUCATION SOCIETY'S PILLAI HOC COLLEGE OF ENGINEERING & TECHNOLOGY RASAYANI TALUKA PANVEL," Unique Plug-In Hybrid Electric Vehicles".
- Dr. Biplab Kumar Sarkar, 202232217362700DE, MAHATMA EDUCATION SOCIETY'S, PILLAI HOC COLLEGE OF ENGINEERING & TECHNOLOGY RASAYANI TALUKA PANVEL, "A DIGITAL VOTING AND FRAUD DETECTION SYSTEM USING BLOCK CHAIN AND MACHINE LEARNING" 2022, Indian Patent Office.
- Dr. Biplab Kumar Sarkar, 20221011560, MAHATMA EDUCATION SOCIETY'S PILLAI HOC COLLEGE OF ENGINEERING & TECHNOLOGY RASAYANI TALUKA PANVEL, "Advanced Method and Process Quick Electric Vehicle Charging." 2022, Indian Patent Office.

Chapter - 12

Impact on Mental Health Using Machine Learning

Dr. Preeti Sharma

Professor, Inderprastha Engineering College, Ghaziabad, (U.P), India.

Komal Desai Goyal

Balaji Institute of Telecom and Management, Sri Balaji University, India. Email: komaldesai214@gmail.com

Prof (Dr.) B.K. Sarkar

MES's, Pillai HOC College of Engineering &Technology,
Rasayani Taluka Panvel, Dist, Navi Mumbai, Maharashtra 410207, India.

ABSTRACT

Our Invention "Impact on Mental Health using Machine Learning Wellbeing of graduate Students due to pandemic" is a days of public health crisis, the psychological capacity of the public to deal with ongoing pandemic is of a global worry and this is not applicable to individuals with COVID19 only but also those who are in quarantine for a long time during a lockdown. The world has witnessed the massive changes in all fronts, and education industry (academia) is no exception. The COVID-19 pandemic has had a significant worldwide impact on the lives of many people especially more than a billion students during the first wave (UNESCO 2021). The American Psychological Association (APA) reports 81% of Gen Z teens have experienced more intense stress during the COVID-19 pandemic. Major technological transformation, agility and disruption have been observed in educational sectors from schools to professional colleges. As of June 2020, almost every nation in world have shut schools and educational institutes across the country, either postponing or prohibiting campus activities. The teachers, students and their families have experienced psychological distress in the form of anxiety, anger, confusion and stress like symptoms that has led to degradation of mental health. The high range of mental consequences in college students have been noticed in different studies conducted worldwide. According to Young's Mind Survey 75% of people, think that they have found the current lockdown harder to cope with than the previous ones including 44% who said it said it was much harder. Above 67% stated that the pandemic will have enduring negative effect on the mental health of students. Almost every student is suffering from anxiety and it is important to explore the mental health and wellbeing of students during this pandemic. Therefore, the proposed research will study to predict the impact of epidemic on Mental Health and Wellbeing among graduate Students using machine learning algorithms. An ensembled predictive model will be developed to find the change in the mental health and to study the factors that have contributed to degrade the mental health and wellbeing of the students.

FIELD OF THE INVENTION

Our Invention is related to a develop an ensemble model using machine learning techniques to study the impact on Mental Health and Wellbeing of graduate Students due to pandemic

BACKGROUND OF THE INVENTION

Define the research problem: The main research questions are – has the mental health of students changed during the COVID-19 pandemic and what factors have contributed to the change in mental health? What are the major challenges faced by students due to home quarantine activities?

What are the social stressors and psychological discomfort faced by the students?. Addressing the mental health of college students and suggesting some preventive strategies have become the need of the hour. This research is also aimed to elicit potential approaches to mitigate the degradation in mental health from the students

PRIOR ART SEARCH

As the research problem is identified and defined the next step is to study the existing research work carried out in the relevant area. The review of literature navigates the research process to the right direction.The related research work from UNESCO, WHO , PUBMED and NCBI will be studied to find out the impact of students mental health in higher education.

The research papers from refered international journals and

conferences will be reviewed to understand the quantitative and qualitative analysis executed in the problem area. Psychiatrists and therapists will be consulted to understand the depth of the problems faced by students and its impact.

OBJECTIVES OF THE INVENTION

- To collect the demographic data and related factors that created an impact among the students during pandemic.
- To identify the challenges faced by students due to Home quarantine activities and related social factors.
- To study various parameters that affects mental health and wellbeing of students during pandemic.
- To develop and fit an ensemble machine learning model to predict and identify the factors that effects on mental health and wellbeing of students during pandemic.
- To suggest the preventive strategies to improve the mental health and wellbeing of students and provide adequate support and guidance.

SUMMARY OF THE INVENTION

Data Collection

The primary data to conduct this research will be done through survey. A questionnaire pertaining to the study will be prepared and collected through interview method, google forms, email, WhatsApp and other social media platforms. The data will be collected from Engineering and Management College Students across Maharashtra State in INDIA. Secondary data will be collected from various articles, websites, and journals and published research papers.

Pilot Study

In this phase preliminary experiments will be performed based on the data collected using the above methods. The knowledge Discovery in Databases (KDD) process will be implemented in machine learning to find the pattern. Various data analysis techniques will be incorporated to test the data and fit the ML model in the pilot study.

AI & ML Model Fitting

Based on the results obtained from the pilot study, the model will be fine-tuned and ensembled ML model will be created and tested with all data to predict the factors impacting the mental health and wellbeing of students in higher education.

BRIEF DESCRIPTION OF THE DIAGRAM

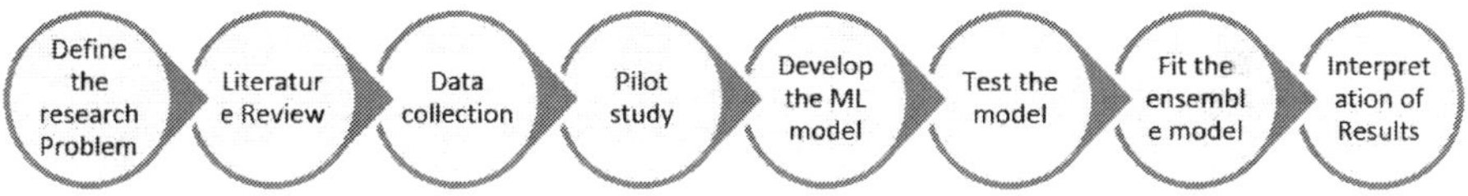

Fig. 1: Methodologies

DESCRIPTION OF THE INVENTION

The COVID-19 pandemic has had a significant worldwide impact on the lives of many people especially more than a billion students during the first wave (UNESCO 2021).

The American Psychological Association (APA) reports 81% of Gen Z teens have experienced more intense stress during the COVID-19 pandemic. Major technological transformation, agility and disruption have been observed in educational sectors from schools to professional colleges.

The outcome of this research will be helpful to monitor the students' mental health and wellbeing not only in this pandemic but also to handle any untoward circumstances in future.

Various factors that are creating negative impact on the mental health of students such as depression, restlessness and anxiety due to the lockdown and other restrictions imposed in the country will be intimated to colleges.

An appeal will be made to higher education institutes to setup a care and counselling center to address these issues immediately and strengthen the mental health of the students.

Various psychosocial factors, family issues, low social support and mental health problems of students discovered in this study should be addressed immediately and necessary support and services to be provided to those students at risk.

The current study indicates timely assessment and a call for continued access to mental health services to ensure graduate students receive adequate support during and after the pandemic.

Overall this research will be a major contribution to the mankind to continuously monitor their wellbeing and guide them to address their psychosocial, economical and mental problems caused due to pandemic.

Report and Interpret Results

After finishing with Model fitting in this stage, the significance of the research study and its outcome will be described. Various Data visualization reports will be generated for the predicted results and the factors impacting the students' academic performance.

The obtained results will be interpreted. Major stressors due to pandemic will be identified and their effects on college students' mental health will be discussed. The results could assist the institutes to provide appropriate support services to enhance students' adjustment and promote.

WE CLAIMS

Our Invention "To develop an ensemble model using machine learning techniques to study the impact on Mental Health and Wellbeing of graduate Students due to pandemic" is a days of public health crisis, the psychological capacity of the public to deal with ongoing pandemic is of a global worry and this is not applicable to individuals with COVID19 only but also those who are in quarantine for a long time during a lockdown. The world has witnessed the massive changes in all fronts, and education industry (academia) is no exception. The COVID-19 pandemic has had a significant worldwide impact on the lives of many people especially more than a billion students during the first wave (UNESCO 2021). The American Psychological Association (APA) reports 81% of Gen Z teens have experienced more intense stress during the COVID-19 pandemic. Major technological transformation, agility and disruption have been observed in educational sectors from schools to professional colleges. As of June 2020, almost every nation in world have shut schools and educational institutes across the country, either postponing or prohibiting campus activities. The teachers, students and their families have experienced psychological distress in the form of anxiety, anger, confusion and stress like symptoms that has led to degradation of mental health. The high range of mental consequences in college students have been noticed in different studies conducted worldwide. According to Young's Mind Survey 75% of people, think that they have found the current lockdown harder to cope with than the previous ones including 44% who said it said it was much harder. Above 67% stated that the pandemic will have enduring negative effect on the mental health of students. Almost every student is suffering from anxiety and it is important to explore the mental health and wellbeing of students during this pandemic. Therefore, the proposed research will study to predict the impact of epidemic on Mental Health and Wellbeing among graduate Students using machine learning algorithms. An ensembled predictive model will be developed to find the change in the mental health and to study the factors that have contributed to degrade the mental health and wellbeing of the students.

- According to claim1# the invention is to a " To develop an ensemble model using machine learning techniques to study the impact on Mental Health and Wellbeing of graduate Students due to pandemic" will identify the stressors that has impacted students psychologically, socially, emotionally and economically.
- According to claim1# the invention is to a provide the information about the risk factors connected with Mental Health and Wellbeing of students of Professional Courses from Maharashtra state, which will be helpful to concerned authorities to plan and adopt appropriate interventions to overcome the negative impacts to ensure sound mental health and wellbeing of Students.
- According to claim1# the invention is to a various indicators highlighted in this study could guide the detection and the counselling of students most at risk in this type of situation. The indicators obtained from this research study in the post pandemic will be helpful to form a counselling centre to care and help the students.
- According to claim1# the invention is to a makers and higher education institutions around the country may benefit from the findings of this work while formulating policy recommendations and strategies to support students during this and any future pandemics.

CITATIONS

- Rohini Bhosale, 202241021998, MAHATMA EDUCATION SOCIETY'S .PILLAI HOC COLLEGE OF ENGINEERING &TECHNOLOGY RASAYANI,TALUKA PANVEL, "IoT-Based Automatic Railway Gate Controller and Notify." 2022, Indian Patent Office,
- Ankit Arun Ulvekar, Juily Dinesh Nimbolkar, Rushikesh Suhas Haladkar, Kunal Sudhakar Gaikwad, Ms. Manisha B. Jamgade, Dr. G R Patil, 202221021777, MAHATMA EDUCATION SOCIETY'S, PILLAI HOC COLLEGE OF ENGINEERING &TECHNOLOGY RASAYAN PANVEL, "Recycled Concrete Using Demolition Waste with Rice Husk and Coir fibres For Sustainability" 2022, Indian Patent Office.
- 202221023844, Ar. Shubhangi Bhide, Mahatma Education Society's Pillai College of Architecture, "ECO FRIENDLY IV & Saline Stand OF Bamboo Material" 2022, Indian Patent Office.
- Mimi Cherian, Satishkumar L. Varma, 202221021778, Mahatma Education Society's, " SDN Enabled Secure Framework for IoT using Dynamic Counter-Based Approach and Deep Learning Model", 2022, Indian Patent Office.
- Ashish Purushottam Kulkarni, Purushottam Gopalrao Kulkarni, 202221024947, MAHATMA EDUCATION SOCIETY'S, PILLAI HOC COLLEGE OF ENGINEERING &TECHNOLOGY RASAYANI, TALUKA PANVEL, DIST, NAVI MUMBAI," Minimization of accidents due to the Extra gap between the Edge of Platform and Footboard of the Train." 2022, Indian Patent Office.
- Meera Ranjith, Neelam Raju Gupta, Gnana Vidya, Jasprit Kaur Kawaljit Singh Gupta, Poorni Raju Swamy
- Dr. B.K. Sarkar,202022102061, Granted, Mahatma Junior College of Education, Opposite Fire Brigade,Sion Trombay Road, Chembur Naka,Mumbai 400071,India, " Teaching Kitchens as learning labs for life skills in secondary school curriculum" Germany Govt
- Dr. B.K .Sarkar,202022101506, Granted, MAHATMA EDUCATION SOCIETY'S, PILLAI HOC COLLEGE OF ENGINEERING &TECHNOLOGY RASAYANI, TALUKA PANVEL, DIST, NAVI MUMBAI, MAHARASHTRA 410207, INDIA. "Digital voting and fraud Detection using Block Chain and Machine Learning System." Germany Govt. MR. SATPUTE SUNNY SHIVAJI, MR.RANJEET KUMAR YADAV, MR.PRITAM AMBO MUNDHE, PROF.ASOKAN SELVARAJ

- 202221025962,MAHATMA EDUCATION SOCIETY'S, PILLAI HOC COLLEGE OF ENGINEERING &TECHNOLOGY RASAYANI, TALUKA PANVEL, "INTELLIGENT SOLAR HYBRID DEHYDRATOR" 2022, Indian Patent Office.
- Mrs. Jyoti Raju Kadam, Mr. Raju A Kadam, Sanjana Raju Kadam202221027979, MAHATMA EDUCATION SOCIETY'S, Assistant Professor, Department Biotechnology, Pillai College of Arts Commerce and Science. (Autonomous), "Intelligent LED Flat Screen Monitor take a Selfies Pictures and Video Recording" 2022, Indian Patent Office.
- Mrs. Jyoti Raju Kadam, Mr. Raju A Kadam, Sanjana Raju Kadam, 202221027981, MAHATMA EDUCATION SOCIETY'S, Assistant Professor, Department Biotechnology, Pillai College of Arts Commerce and Science. (Autonomous), "Smartphones Convert in to a Mini Projector Lens of High Definition" 2022, Indian Patent Office.
- Mrs. Jyoti Raju Kadam, Mr. Raju A Kadam, 202221027980, MAHATMA EDUCATION SOCIETY'S, 2022, Indian Patent Office. Department Biotechnology, Pillai College of Arts Commerce and Science. (Autonomous), "Built-in Mini Fan of 1 Inch Diameter under Smartphones" 2022, Indian Patent Office.
- Dr. Farhat Fatma Mumtaz Husain Shaikh, Sunita Charanjit Saini, Nikhat Fatma Mumtaz Husain Shaikh 202221028002, MAHATMA EDUCATION SOCIETY'S, Chembur Mumbai, A 302, Neelgagan CHS, Plot No. 33, Sector 42, Seawoods, Nerul West, Navi Mumbai 400706, " Slide Lipstick Holder with Button." 2022, Indian Patent Office.
- Mahatma Education Society's, Prof. Dr. Biplab Kumar Sarkar, 202211028005, Mahatma Education Society's Pillai HOC College Of Engineering &Technology Rasayani, Taluka Panvel, Dist, Navi Mumbai, Maharashtra 410207, India. "-UV- Light Box: Modified Version of UV- Sanitizers", 2022, Indian Patent Office.
- Mahatma Education Society, 202241028431, Mahatma Education Society, Chembur Naka, Mumbai - 400 071, Maharashtra, India." Decresed the Search engine Time Cloud Data based using Machine Lerning" 2022, Indian Patent Office.

CHAPTER - 13

Environmental Impacts Study of Residential Building Using Machine Learning

Dr. Anuj Kumar Sharma

Associate Professor, Civil Engineering Department,
Ramdeobaba College of Engineering and Management Nagpur 13

Mange Amar Vilasrao

MES's, Pillai HOC College of Engineering & Technology,
Rasayani Taluka Panvel, Dist, Navi Mumbai, Maharashtra 410207, India

Sumit S Dalvi

MES's, Pillai HOC College of Engineering &Technology,
Rasayani Taluka Panvel, Dist, Navi Mumbai, Maharashtra 410207, India

ABSTRACT

Our Invention "Environmental Impacts study of Residential Building using Machine Learning" is a construction is considered as one of the main source of environmental pollution. The demand for construction is increasing day by day in the world. The main aim of this study is to evaluate the most common environmental impacts of residential building during construction. The study will help to identify the most common environmental impacts according to their impact level, frequency and consequences of occurrence. This study will help the organizations and project manager or construction manager to increase the knowledge of people which are involved in construction through training and awareness program. The aim of this research is to study implementation of EMS (Energy Management System) in construction companies and to improve the organizational environmental performance. Different methodologies are able to compare the overall impact of various construction projects and rank the significance of various environmental impacts on each project. The environmental impact at particular site is identified primarily at construction stage and their significant impacts are highlighted. Thus, it is possible to implement measures for mitigating adverse impacts, which can then be implemented during construction activities. This research will serve as an assessment tool for the construction project to measure the inputs of their construction activities.

FIELD OF THE INVENTION

Our Invention is related to a PREDICTION OF SEVERE ENVIRONMENTAL IMPACTS RELATED TO CONSTRUCTION PROCESS OF RESIDENTIAL BUILDING.

BACKGROUND OF THE INVENTION

The aim of this research is to study implementation of EMS (Energy Management System) in construction companies and to improve the organizational environmental performance. Different methodologies are able to compare the overall impact of various construction projects and rank the significance of various environmental impacts on each project.

The environmental impact at particular site is identified primarily at construction stage and their significant impacts are highlighted. Thus, it is possible to implement measures for mitigating adverse impacts, which can then be implemented during construction activities. This research will serve as an assessment tool for the construction project to measure the inputs of their construction activities.

OBJECTIVES OF THE INVENTION

1. To identify environmental impacts parameters related to construction process of residential building.
2. To select region for study purpose.
3. To collect data of selected region.
4. To select appropriate software to analyze data.
5. To determine severe environmental impact parameter during construction of residential building.

SUMMARY OF THE INVENTION

A questionnaire is prepared for evaluating the environmental impacts across construction of residential building in selected region depend on Five-point Likert Scale. A questionnaire depends on the probability of occurrence and its severity of impacts on the environment.

A questionnaire depends on site conditions during construction process of residential buildings. In these questionnaire various environmental impacts that occurred during construction process is considered and according to that a questionnaire is prepared.

This questionnaire includes main three environmental impact parameters related to natural resource, ecosystem impact and public impact. In pollution related questionnaire some questions related to air pollution, noise pollution, water pollution, land pollution and dust pollution are included.

In transportation impact related questionnaire some questions related to transportation resources, road traffic due to construction, mode of transport, delay due to transport and rescheduling of activities are included. In resource consumption related questionnaire electricity consumption, water consumption, raw material consumption and sources of water are included.

In public related questionnaire some questions related to public such as site conditions, social disruption issues and public health effects are included.

BRIEF DESCRIPTION OF THE DIAGRAM

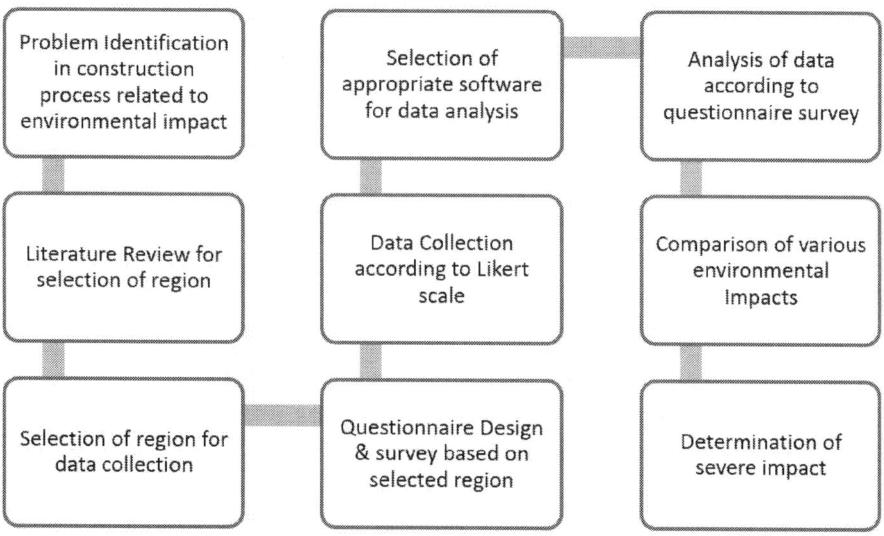

Fig. 1: Methodologies

DESCRIPTION OF THE INVENTION

A Likert scale is also known as psychometric scale. It is commonly used in research questionnaires. The term is often used interchangeably with rating scale, or more accurately the Likert-type scale, even though the two are not synonymous. Likert distinguished between a scale proper, which emerges from collective responses to a set of items (usually eight or more), and the format in which responses are scored along a range.

A survey is conducted to investigate the frequency and severity of environmental impacts across the construction of residential buildings. In this survey five point Likert Scale is used for calculating the rate of frequency of occurrence and severity of environmental impacts.

After this questionnaire survey will collect data. By using appropriate software we will compare & identify most important factors that causes various environmental impacts such as pollution, transportation, resource consumption and pubic impact related. The questionnaire will be filled by experienced site engineers, project managers, project engineers and designers. Following steps are included for analysing the data.

Calculating the impact level by multiplying frequency of occurrence into severity of impact. Impact level is calculating by taking values from data collection. From data collection sheet mean is calculated for frequency of occurrence and severity of impact. Risk factor is calculated by multiplying frequency of occurrence and severity of impact.

For maximum value of impact level first ranking is given. After that ascending order of ranking of factors are given in each category based on impact level.

Calculating the Relative Importance Index (RII)

For more analysis, time series graphs are plotted for frequency of occurrence and severity of impacts for all sites such as small sites and large sites by using Minitab software. Time series graphs are plotted for each category and factor such as for frequency of occurrence and severity of impact.

From time series graphs analysis, it is found that frequency of occurrence and severity of impacts for small sites are more as compared to frequency and severity for large sites.

For more accurate calculations priority matrix for small and large sites are carried out.

For checking data is significant or not, Chi-square test for small and large sites arc carried out by using appropriate software.

OUTCOME

Our Invention "PREDICTION OF SEVERE ENVIRONMENTAL IMPACTS RELATED TO CONSTRUCTION PROCESS OF RESIDENTIAL BUILDING" is a construction is considered as one of the main source of environmental pollution. The demand for construction is increasing day by day in the world. The main aim of this study is to evaluate the most common environmental impacts of residential building during construction. The study will help to identify the most common environmental impacts according to their impact level, frequency and consequences of occurrence. This study will help the organizations and project manager or construction manager to increase the knowledge of people which are involved in construction through training and awareness program. The aim of this research is to study implementation of EMS (Energy Management System) in construction companies and to improve the organizational environmental performance. Different methodologies are able to compare the overall impact of various construction projects and rank the significance of various environmental impacts on each project. The environmental impact at particular site is identified primarily at construction stage and their significant impacts are highlighted. Thus, it is possible to implement measures for mitigating adverse impacts, which can then be implemented during construction activities. This research will serve as an assessment tool for the construction project to measure the inputs of their construction activities.

- According to claim1# the invention is to a help the organizations and project manager or construction manager to increase the knowledge of people which are involved in construction through training and awareness program.
- According to claim1,2# the invention is to a research will serve as an assessment tool for the construction project to measure the inputs of their construction activities.

CITATIONS

- Mahatma Education Society, 202221029221, Mahatma Education Society, Chembur Naka, Mumbai - 400 071, Maharashtra, India. "Raster- Assistive AI based Book Reading Tool" 2022, Indian Patent Office.
- Mahatma Education Society, 202221028885, Mahatma Education Society, Chembur Naka, Mumbai - 400 071, Maharashtra, India." Increased the Soil Stability using W-Plastic." 2022, Indian Patent Office.
- Mahatma Education Society, 202221029340, Mahatma Education Society, Chembur Naka, Mumbai - 400 071, Maharashtra, India." Waste Kitchen and Plastic Convert into an Energy." 2022, Indian Patent Office.
- Mahatma Education Society, 202221029344, Mahatma Education Society, Chembur Naka, Mumbai - 400 071, Maharashtra, India." Method and Process to Improve the Basmati Rice Production", 2022, Indian Patent Office.
- Mahatma Education Society, 202241029342, Mahatma Education Society, Chembur Naka, Mumbai - 400 071, Maharashtra, India." Wireless Material Handling Management Method and Process", 2022, Indian Patent Office.
- Mahatma Education Society, 202221029339, Mahatma Education Society, Chembur Naka, Mumbai - 400 071, Maharashtra, India. "Design and Develop Alarm clock using AI and Cloud Data Based" 2022, Indian Patent Office.
- Mahatma Education Society, 202221029338, Mahatma Education Society, Chembur Naka, Mumbai - 400 071, Maharashtra, India, "Density Based Traffic Signals using Microcontroller using IoT- Based.", 2022, Indian Patent Office.

Made in the USA
Las Vegas, NV
19 September 2026

f9145a8b-aa27-4c18-acdb-334a79585a82R01